Springer Series in

ADVANCED MICROELECTRONICS 4

Springer
Berlin
Heidelberg
New York
Barcelona
Hong Kong
London
Milan
Paris
Singapore
Tokyo

Springer Series in
ADVANCED MICROELECTRONICS

Series editors: K. Itoh, T. Sakurai

The Springer Series in Advanced Microelectronics provides systematic information on all the topics relevant for the design, processing, and manufacturing of microelectronic devices. The books, each prepared by leading researchers or engineers in their fields, cover the basic and advanced aspects of topics such as wafer processing, materials, device design, device technologies, circuit design, VLSI implementation, and subsystem technology. The series forms a bridge between physics and engineering and the volumes will appeal to practicing engineers as well as research scientists.

1 **Cellular Neural Networks**
Chaos, Complexity and VLSI Processing
By G. Manganaro, P. Arena, and L. Fortuna

2 **Technology of Integrated Circuits**
By D. Widmann, H. Mader, and H. Friedrich

3 **Ferroelectric Memories**
By J. F. Scott

4 **Microwave Resonators and Filters for Wireless Communication**
Theory, Design and Application
By M. Makimoto and S. Yamashita

5 **VLSI Memory Chip Design**
By K. Itoh

Series homepage – http://www.springer.de/phys/books/ssam/

M. Makimoto S. Yamashita

Microwave Resonators and Filters for Wireless Communication

Theory, Design and Application

With 161 Figures

Springer

Dr. Mitsuo Makimoto
Dr. Sadahiko Yamashita
Matsushita Research Institute Tokyo, Inc.
3-10-1 Higashimita, Tama-ku
Kawasaki 214
Japan

Series Editors:

Dr. Kiyoo Itoh
Hitachi Ltd., Central Research Laboratory
1-280 Higashi-Koigakubo
Kokubunji-shi
Tokyo 185-8601
Japan

Professor Takayasu Sakurai
Center for Collaborative Research
University of Tokyo
7-22-1 Roppongi, Minato-ku,
Tokyo 106-8558
Japan

Library of Congress Cataloging-in-Publication Data

Makimoto, M. (Mitsuo), 1944-
Microwave resonators and filters for wireless communication : theory, design, and application / M. Makimoto, S. Yamashita.
p. cm. -- (Springer series in advanced microelectronics ; 4)
Includes bibliographical references and index.
ISBN 3540675353 (alk. paper)
1. Microwave filters. 2. Resonators. 3. Wireless communication systems--Equipment and supplies. I. Yamashita, S. (Sadahiko), 1940- II. Title. III. Series.

TK7872.F5 M32 2000
621.3845--dc21

00-041924

ISSN 1437-0387
ISBN 3-540-67535-3 Springer-Verlag Berlin Heidelberg New York

Springer-Verlag Berlin Heidelberg New York
a member of BertelsmannSpringer Science+Business Media GmbH

Typesetting: Data conversion by Satztechnik Katharina Steingraeber, Heidelberg
Cover concept by eStudio Calmar Steinen using a background picture from Photo Studio "SONO". Courtesy of Mr. Yukio Sono, 3-18-4 Uchi-Kanda, Chiyoda-ku, Tokyo
Cover design: *design & production* GmbH, Heidelberg

Printed on acid-free paper SPIN: 10754512 57/3141/mf 5 4 3 2 1 0

Preface

Anytime. Anywhere.
Mobile communication lets you keep in touch.

The introduction of radio wave transmission by G.M. Marconi and the telephone by Graham Bell in the late 19th century together brought dramatic changes in the way we communicate. By combining the two, we now possess a powerful communication tool enabling us to keep in touch with anyone, anytime and anywhere. Technological breakthroughs have contributed to this scheme by providing compact electronic devices, and without such progress, our present situation could not have existed. We foresee a bright future based on the further progress of mobile communication. The authors of this book have engaged in the downsizing of radio frequency (RF) circuits found in wireless communication devices. Filtering devices are an essential component for such circuits, and considerable effort has been put into the research and development of such devices. This book starts from a conceptual view of such filters, and expands on this idea to provide a practical solution for the application of filtering devices.

The purpose of this book is (1) to provide general information and basic design procedures for filters applied to wireless communication systems, (2) to illustrate the availability and introduce actual design examples of the stepped-impedance resonator (SIR) structure intended for the RF/microwave region, and (3) to propose a more general concept for transmission-line resonators based on an expanded SIR structure. Basic theory and analysis methods for RF/microwave transmission line resonators are extensively explained, followed by the synthesis theory and practical design techniques for filters applying such resonators. Various design examples are also presented in each part. The book is sure to offer useful information for students engaged in microwave circuit theory, researchers in the field of electromagnetic wave engineering, and R&D engineers of wireless communication systems and circuits design.

The features of this book are as follows: (1) The concept of SIR is introduced in order to expand the applicable frequency range of the conventional transmission-line resonator. (2) Novel resonator/filter structures applying newly developed dielectric and superconducting materials along with advanced micromachining fabrication technologies are introduced. (3) Resonator applications are extended beyond filtering devices, to RF devices with new structures such as a balanced mixer and a low phase noise oscillator. Such topics have received scant attention up to now, and thus we are convinced that this book will provide useful information for the present and future development of compact and high performance radio equipment.

The book is organized into six chapters and an appendix. Chap. 1 starts with an overview, including historical perspectives of filters, applied to communication systems. Next, the required functions and characteristics of the various types of filters applicable to wireless communication equipment are presented. Here we introduce the stepped inpedance resonator (SIR) as a promising resonator structure capable of overcoming the essential drawbacks of the conventional transmission-line resonator, which is commonly applied to filters in the RF/microwave region. In Chap. 2, the basic structure and properties of the SIR are analyzed by introducing the concept of transmission-line impedance ratio. Chap. 3 explores the quarter-wavelength type SIRs and their application to filters. Practical use of this SIR has progressed in the shape of coaxial dielectric resonators, and various examples focusing on miniaturization techniques are presented in this chapter. Chap. 4 concentrates on half-wavelength type SIRs and their application. This SIR structure is easily realized with stripline configurations and possesses a high affinity with active devices. Utilizing these advantages, practical applications to mixers and oscillators, as well as filters, are illustrated through experimental examples. Chap. 5 discusses one-wavelength type SIRs. This SIR is usually designed as a dual-mode resonator because it is too large for single mode application. The excitation method and theoretical analysis of the dual-mode resonator is introduced, and a design method for dual-mode filters is derived. Experimental examples are also presented. Finally, in Chap. 6 the wider concept and technological trends of SIRs are presented. Here, topics such as expansion of applicable frequency range, problems to be solved, and promising applications of SIRs are discussed and summarized. In addition, in the Appendix we introduce the method of analysing of single resonators and resonator-pairs using a general-purpose microwave circuit simulator. This analysis method provides a generalized estimation of resonator properties for filter synthesis, without having to develop an application-specific computer program. This technique is applied to filter design throughout this book, and can be recommended as a highly practical design method.

As previously mentioned, it seems evident that wireless communications will progress in the direction of global personal communication systems based on multimedia content. Accordingly, the role of the filtering device as a key RF component will become even more important, and requirements for performance will become more critical. We hope this book will contribute to the design and development of filters for wireless communications, and promote further development of advanced filtering devices.

Most of the technical material relating to SIRs in this book is compiled from R&D results achieved at the Matsushita Research Institute Tokyo in the past 25 years. We would like to express our sincere gratitude to the many researchers who have travelled a long way to engage in this project, along with our colleagues at Matsushita Communication Industrial Co. who have provided us with valuable advice and evaluation leading to the achievement of publishing this book. We would also like to deeply thank Dr. M. Sagawa, Dr. H. Yabuki, Mr. M. Matsuo, and Dr. A. Enokihara for their courtesy in offering us valuable experimental data on filter establishment.

Finally, we would like to thank Mr. K. Goho who put much effort into translating this book, as well as Dr. C. Ascheron of Springer-Verlag for providing us the opportunity and helpful advice in publishing this book.

Kawasaki, Japan
Summer, 2000

Mitsuo Makimoto
Sadahiko Yamashita

Contents

1. Introduction ... 1
1.1 The History of Filters in Telecommunication ... 1
1.2 Filters for Wireless Communication ... 3
1.3 Classification of Resonators and Filters for Wireless Communication ... 5
1.4 Transmission-Line Resonators and Stepped Impedance Resonator (SIR) ... 7

2. Basic Structure and Characteristics of SIR ... 11
2.1 Basic Structure of SIR ... 11
2.2 Resonance Conditions and Resonator Electrical Length ... 12
2.3 Spurious Resonance Frequencies ... 16
2.4 Derivation of an Equivalent Circuit of SIR ... 16

3. Quarter-Wavelength-Type SIR ... 19
3.1 Analysis of $\lambda_g/4$-Type Coaxial SIR ... 20
3.1.1 Impedance Ratio R_Z ... 20
3.1.2 Effects of Discontinuity ... 21
3.1.3 Unloaded-Q of Coaxial SIR ... 24
3.2 Bandpass Filters Using Coaxial SIR ... 30
3.2.1 Synthesis Method of SIR-BPF Using Capacitive Coupling ... 30
3.2.2 Design Examples and Performances ... 32
3.3 Double Coaxial SIR (DC-SIR) ... 39
3.3.1 Advantages of DC-SIR ... 39
3.3.2 Resonance Condition and Unloaded-Q ... 39
3.3.3 400 MHz-Band High-Power Antenna Duplexer ... 41
3.4 Dielectric Coaxial SIR ... 45
3.4.1 Dielectric Materials and Features of Dielectric Resonators ... 45
3.4.2 Basic Structure and Characteristics of Dielectric Coaxial Resonator ... 47
3.4.3 Design Example of Antenna Duplexer for Portable Radio Telephone ... 50
3.4.4 Dielectric DC-SIR ... 53
3.4.5 Dielectric Monoblock SIR-BPF ... 58

3.5 Stripline SIR ... 60
3.5.1 Basic Structures and Features ... 60
3.5.2 Coupling Between Resonators ... 61
3.5.3 Stripline SIR-BPF ... 62

4. Half-Wavelength-Type SIR ... 65
4.1 Stripline $\lambda_g/2$ Type SIR ... 66
4.1.1 Basic Characteristics ... 66
4.1.2 Equivalent Expressions for Parallel Coupled-Lines Using Inverter ... 67
4.1.3 Synthesis of Stripline Parallel-Coupled SIR-BPF ... 71
4.1.4 Filter Design Examples ... 74
4.2 Internally Coupled SIR ... 84
4.2.1 Basic Structures and Resonance Condition ... 84
4.2.2 Equivalent Circuits at Resonance ... 89
4.2.3 Filter Design Examples ... 91
4.2.4 Application to Oscillator and Mixer Circuits ... 101

5. One-Wavelength-Type SIR ... 107
5.1 Orthogonal Resonance Modes in the Ring Resonator ... 107
5.2 Application of λ_g-Type SIR as Four-Port Devices ... 109
5.3 Application of λ_g-Type SIR as Two-Port Devices ... 112
5.3.1 Coupling Means for Orthogonal Resonant Modes ... 112
5.3.2 Analysis of Coupling Between Orthogonal Resonant Modes ... 115
5.3.3 Application to Filters ... 119

6. Expanded Concept and Technological Trends in SIR ... 127
6.1 SIR Composed of Composite Materials ... 127
6.1.1 Combination of Magnetic and Dielectric Materials ... 127
6.1.2 Coaxial SIR Partially Loaded with Dielectric Material 128
6.2 Multistep SIR and Tapered-Line Resonators ... 131
6.3 Folded-Line SIR ... 139
6.4 Technological Trends of SIR in the Future ... 145

Appendix. Analysis of Resonator Properties Using General-Purpose Microwave Simulator ... 149
A.1 Design Parameters of Direct-Coupled Resonator BPF ... 149
A.2 Filter Design by Experimental Method ... 150
A.3 Determination of Q and k Using General Purpose Microwave Simulator ... 151
A.3.1 Determination of Q ... 151
A.3.2 Determination of Coupling Coefficient ... 154

References ... 157

Index ... 161

1. Introduction

1.1 The History of Filters in Telecommunication

Filters in electric circuits have played an important role since the early stages of telecommunication, and have progressed steadily in accordance with the advancement of communication technology. In 1910, the introduction of the carrier telephony system – a novel transmission system of multiplex communication – drastically reformed the technological landscape surrounding telecommunications and introduced a new era in telecommunications. The system required the development of new technology to extract and detect signals contained within a specific frequency band, and this technological advance further accelerated the research and development of filter technology.

In 1915, the German scientist K.W. Wagner introduced a filter design method which became well known as the "Wagner filter". Meanwhile, another design technique was under development in the United States by G.A. Cambell, a design which later became known as the image parameter method. After these technological breakthroughs, many notable researchers including O.J. Zobel, R.M. Foster, W. Cauer and E.L. Norton actively and systematically studied filter design theory using lumped element inductors and capacitors. Subsequently, a precise filter design method with two specific design steps was introduced in 1940. The first step in this filter design method was the determination of a transfer function that met with the required specifications. Then, using a frequency response estimated by the previous transfer function, the second step was to synthesize electrical circuits. The efficiency and success of this filter design method were unmatched, and most current filter design techniques today are based on this early method.

Soon, filter design development expanded from the lumped-element LC resonators to the newly discovered field of distributed-element coaxial resonators or waveguide resonators [1]. Simultaneously, broad advances in the field of filter materials were achieved, greatly advancing the progress of filter devices. In 1939, P.D. Richtmeyer reported that the dielectric resonator [2], which utilized electromagnetic wave resonance, had two special features: small size and high Q value. However, the material's lack of temperature stability in those days meant that the filter was insufficient for practical use. In the 1970s, the development of various kinds of ceramic materials with excellent temperature stability and high Q value increased the viability of practical application of the dielectric filter [3]. With the development of these ceramic materials, application to filters advanced rapidly. The dielectric filter has since become one of the most important and familiar components in recent RF/microwave communication equipment. In addition, superconducting materials with high critical temperature discovered in the 1980s are expected to have the possibility of designing novel microwave filters with extremely low loss and small size [4]. Many R&D efforts have been directed toward developing a practical use.

In the early stages of filter development, filter design concentrated on passive electrical circuits composed of appropriate combinations of inductor L and capacitor C. The LC resonator being a linear resonant system, many early researchers believed that resonant systems based on physical principles other than lumped/distributed-element electrical circuits could achieve filter performance. In 1933, W.P. Mason revealed a quartz crystal resonator filter [5], and this filter soon became an indispensable component in communication equipment because of its excellent temperature stability and low loss characteristics. Like the crystal resonator, the ceramic resonator system uses bulk waves. Although the ceramic filter does not offer many of the valuable properties of the crystal filter, they are often utilized because of their low production cost. Surface acoustic wave (SAW) resonators which employ single crystal material such as $LiNbO_3$, $LiTaO_3$, etc. can also be used as filter elements, and SAW filters are feasible for practical use at much higher frequency ranges than bulk-wave filters [6]. A resonant system of magnetostatic modes, generated by applying a biasing magnetic field to a ferrite single crystal, also has the capability to serve as a filter [7]. Microwave filters using a YIG (yttrium iron garnet) sphere similarly have been put to practical use. The YIG filter's special feature is the ability to change the center frequency by varying the magnetic field strength.

Although all the above mentioned filters utilize a linear resonant system, early stages of filter development suggested the probable and likely existence of another approach that would realize filter response. The main reason behind this belief was the general view of filters as functional devices which achieved their performances according to given transfer functions. Filtering devices which employed active circuits were typical examples. In the "vacuum tube" age, active RC filters without LC reactance circuits were aggressively studied and developed, and the results of this research have also been utilized in filter technology. Some examples of such active filters include a popular technique that establishes equivalent LC circuits through the use of gyrators, and a technique that realizes the transfer function response by using operational amplifiers with feedback circuits. The advancement of the semiconductor analog-integrated circuits stimulated and promoted the progress, practical use, and spread of these active filters.

In addition to the previously described techniques, there is also a digital technique [8] which realizes the transfer function of a filter more directly.

The general procedure for this technique is: first the input analog signal is converted into a digital signal; next numeric operation processing is performed according to the transfer function, and finally the output signal is obtained after digital-to- analog conversion. Although the theory that such digital filters exist was proposed much earlier, industrial applications did not materialize until the remarkable digital LSI advancements in the 1970s. Currently, almost all digital communications systems adopt these digital filters as the base-band filters. In addition, the progress of hardware and high-speed operation algorithms continually extend the upper bounds of applicable frequency.

As described above, filters and their design methods have a long history. Filters have become indispensable devices not only in the field of telecommunication, but also in many other types of electrical equipment. Due to the variety and diversity of filter types, it often becomes necessary for a designer to carefully consider which filter to adopt for a particular application.

When classifying filters according to frequency response, they are divided into four basic types: LPF (low-pass filter), HPF (high-pass filter), BPF (bandpass filter) and BEF (band-elimination filter). The LPF design method provides the basis for all types of filters. Many tutorial books describe how the design techniques for HPF, BPF and BEF can be derived from the LPF prototype design method by appropriate frequency conversion. This book concentrates mainly on BPF, because BPFs remain the most important and complicated to design for use in wireless communication equipment.

1.2 Filters for Wireless Communication

Various kinds of filters are used for wireless communication equipment. In this chapter, the type of filter, its function, and the necessity of filter miniaturization will be addressed, beginning with one of the most familiar forms of wireless communication equipment: the mobile telephone terminal.

Figure. 1.1 shows a typical RF circuit block diagram of a mobile telephone based on a FDMA-FDD system (frequency division multiple access-frequency division duplex,) the generally adopted system for first generation mobile communication [9]. The receiver component employs a double super heterodyne system in this example, and the received signal from the antenna is amplified by a low noise amplifier after undesired signals are removed by the Rx-BPF, a part of the duplexer. This signal is then transferred through the quartz crystal BPF and to the IF port after frequency conversion by the mixer. The quartz crystal BPF plays the role of a channel filter and selects only one specific channel signal among frequency multiplexed signals. This IF signal is again amplified and converted to the second IF signal, and finally becomes the base-band signal after passing through the detector and demodulator.

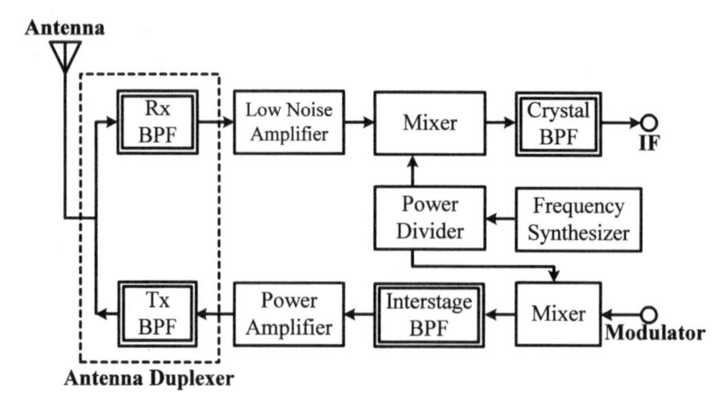

Fig. 1.1. A typical RF circuit block diagram of a mobile telephone

The transmitter section in Fig. 1.1 adopts an up-conversion system which up-converts a premodulated IF signal. Because the output signal of the mixer contains various spurious signals, the desired signal must be extracted with a BPF and then amplified by the power amplifier. The power amplifier is another source of noise, where spurious signals, primarily harmonic frequency components of input signals, are generated. These undesired signals are removed by the Tx-BPF, a part of the duplexer, and finally the signal is radiated from the antenna in the form of an electromagnetic wave.

Therefore, filters are of immeasurable importance to wireless equipment because they are always located between the input and output stages of the RF active circuits. Next, the functions and the required characteristics of each individual filter will be addressed in detail.

The necessary functions of the Rx-BPF are (1) avoiding saturation of the receiver front-end due to leakage of the output signal from the transmitter, (2) removal of interference signals such as the image frequency signal component, and (3) reduction of power leakage of the local oscillator from the antenna. Therefore, the optimal performance of the Rx-BPF would achieve high attenuation for removal of interference and simultaneously reduce pass-band insertion loss that directly affects the sensitivity of the receiver. The primary function of the crystal BPF is channel selection, therefore it must hold properties such as steep attenuation, good group delay characteristics, and excellent temperature stability as a narrow band filter.

The primary function of the Tx-BPF is the reduction of spurious radiation power from the transmitter to avoid interference with other wireless systems. The dominant frequency components of these undesired signals are the second or third harmonics of the transmitting signal frequency, as well as the local oscillator frequency. Another vital function is the attenuation of noise within the transmission signal at the Rx band and thus suppression of its level below the sensitivity point of the receiver. Thus, the Tx-BPF must possess a wide stop band for spurious signal suppression while maintaining a low pass-band insertion loss and the capability of handling large signal levels at the output stage.

From a practical point of view, miniaturization is an important issue for all portable electronic equipment. The reduction of size and weight becomes especially critical for wearable equipment such as portable mobile phones. Obviously, there have been great demands and expectations for the miniaturization of RF circuits. Almost all circuits of lower frequency in the IF and baseband section, including filters, can employ digital LSIs (large scale integrated circuit). Therefore, the miniaturization of these circuits will progress along with the advancement of semiconductor technology.

On the contrary, the introduction of MMICs (monolithic microwave integrated circuit) suggests the miniaturization of RF active circuits such as amplifiers, modulators and frequency converters will be increasingly possible. Still, there are a number of problems to be solved before obtaining optimal circuit size reduction in circuits that possess resonators such as RF filters and oscillators. Therefore, size reduction and achieving high performance of filters continue to be important research issues, and will likely trigger a new technology frontier in this field in the areas of circuit theory, materials, minute processing technology, and rigorous design techniques.

1.3 Classification of Resonators and Filters for Wireless Communication

The main frequency bands assigned to wireless communication are spread throughout a wide range, from several tens of MHz to several tens of GHz. Since a wide variety of resonators and filters can be applied to these frequency bands, Fig. 1.2 shows the applicable range of some typical examples. Figure 1.2 omits active circuit filters such as active RC filters, SCFs (switched capacitor filters) and digital filters because such microwave application filters are not feasible at the present time and exist only at an R&D stage. Furthermore, Fig. 1.2 also omits magnetostatic mode filters, which have been applied mainly to measuring instruments, because of their rare use in wireless communication equipment.

For frequencies below 1 GHz, the figure suggests the most commonly used resonators are the bulk-wave, SAW and helical resonators. Since each resonator type has its own advantages and disadvantages, it is necessary to adopt the proper filter according to the application and purpose of the filter. Bulk-wave and SAW resonators/filters are applied in cases where there is strong demand for miniaturization and low-loss characteristics; helical resonators/filters are often utilized when a high level of power handling is necessary. In addition, bulk-wave and SAW resonators show outstanding temperature characteristics, thus satisfying conditions for application to narrow band filters.

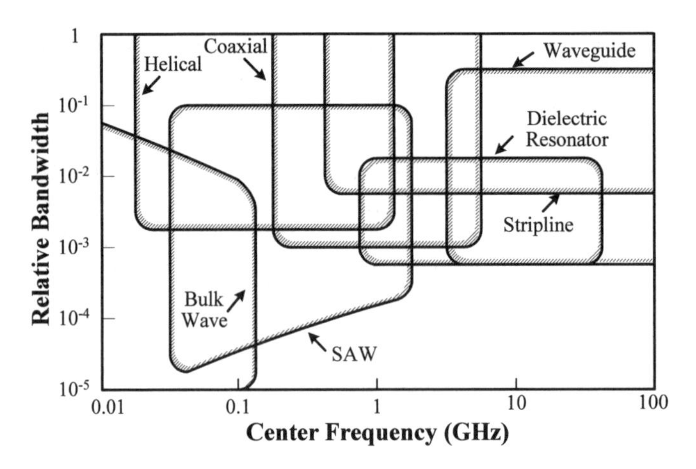

Fig. 1.2. Applicable frequency range of typical resonators and filters

For a frequency range from RF to microwave, various kinds of resonators including the coaxial, dielectric, waveguide, and stripline resonators exist [1]. Coaxial resonators have many attractive features including an electromagnetic shielding structure, low-loss characteristics and a small size, but their minute physical dimensions for applications above 10 GHz make it difficult to achieve manufacturing accuracy. Dielectric resonators also possess a number of advantages such as low-loss characteristics, acceptable temperature stability and a small size. However, high cost and present-day processing technology restrictions limit dielectric resonator utilization to applications below 50 GHz.

Waveguide resonators have long been used in this frequency range, possessing two main advantages: low-loss characteristics and practical application feasibility up to 100 GHz. However, the greatest drawback of the waveguide resonator is its size, which is significantly larger than other resonators available in the microwave region.

Presently, the most common choice for RF and microwave circuits remains the stripline resonator. Due to practical features including a small size, easy processing by photolithography, and good affinity with active circuit elements, many circuits utilize the stripline resonator. Another advantage of the stripline resonator is a wide applicable frequency range which can be obtained by employing various kinds of substrate materials. However, a major drawback to the use of the resonator is a drastic increase in insertion loss compared to other types of resonators, making it difficult to apply such stripline resonators to narrow band filters. Still, such stripline resonators yield high expectations for application to ultra low-loss superconducting filters [4], which are now under development and require fabrication methods using planar circuits such as stripline configuration.

1.4 Transmission-Line Resonators and Stepped Impedance Resonator (SIR)

The most typical transmission-line resonators utilizing transverse electromagnetic modes (TEM) or quasi-TEM modes are coaxial resonators and stripline resonators. As shown in Fig. 1.2, these resonators possess a wide applicable frequency range starting at several 100 MHz extending to arround 100 GHz, and presently remain the most common choice for filters in wireless communication [10]. These resonators, as previously described, do not possess low-loss properties, i.e., they do not have high Q values compared to waveguide or dielectric resonators. However, they do have valuable features as electromagnetic wave filters: a simple structure, a small size, and the capability of wide application to various devices. Moreover, the most attractive feature of micro-stripline, stripline or coplanar-line resonators is that they can be easily integrated with active circuits such as MMICs, because they are manufactured by photolithography of metalized film on a dielectric substrate.

Figure 1.3 shows the fundamental structure of a micro-stripline half-wavelength resonator with two open-circuited ends, illustrated as a typical example of a transmission line resonator most commonly used in the microwave region. The figure shows the physical structure of the resonator: a strip conductor of uniform width and an overall length equivalent to half-wavelength, formed on a dielectric substrate. This structure can be expressed in electrical parameters as a transmission line possessing uniform characteristic impedance with an electrical length of π radian. Such transmission-line resonators will be referred to as uniform impedance resonator (UIR). General requirements for UIR intended dielectric substrate materials include a low loss-tangent, high permitivity, and temperature stability. Transmission-line resonators are widely used because of their simple structure and easy-to-design features, as illustrated in this example, and many scientists feel that design methods of conventional filters using such UIRs have been perfected. In practical design, however, such resonators have a number of intrinsic disadvantages, such as limited design parameters due (ironically) to their simple structure. Other electrical drawbacks include spurious responses at integer multiples of the fundamental resonance frequency.

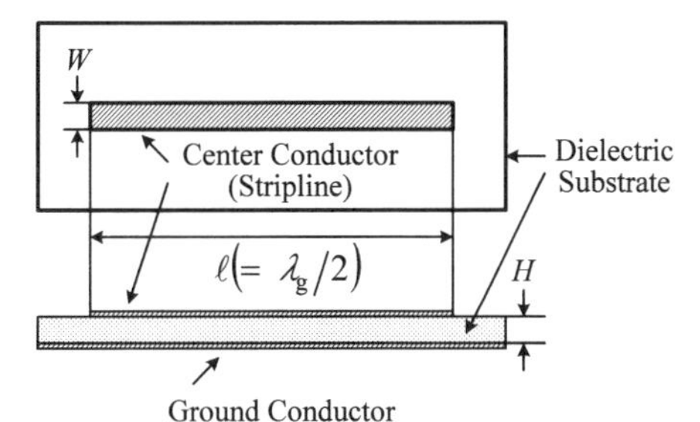

Fig. 1.3. Fundamental structure of a micro-stripline half-wavelength resonator

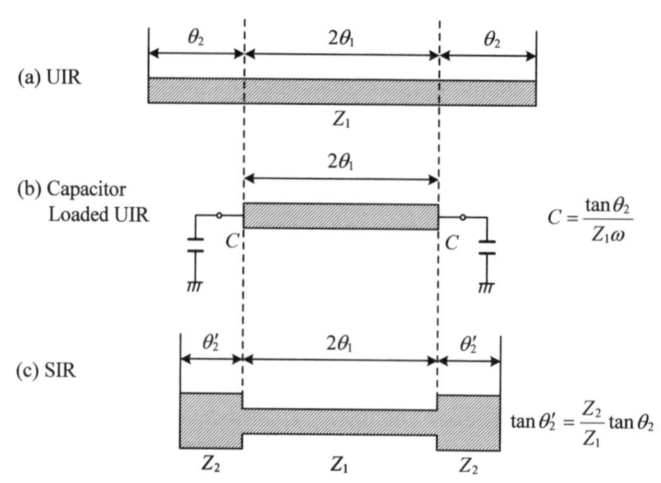

Fig. 1.4. Structural variations of a half-wavelength type resonator. (**a**) Uniform impedance resonator (UIR). (**b**) Capacitor loaded UIR. (**c**) Stepped- impedance resonator (SIR)

To overcome these problems, it is a common practice in the VHF band to load capacitors at both open-ends of the resonator. By doing so, the resonator length is shortened and spurious resonance frequencies are consequently shifted from the integer multiples of its fundamental frequency.

Figure 1.4 shows the structural variations of a half-wavelength type resonator. The capacitor loaded UIR shown in the figure has a characteristic impedance of Z_1 and an electrical length of $2\theta_1$. When the angular resonance frequency ω_0 of this resonator corresponds to that of a half-wavelength UIR shown in (a), the loading capacitance C is expressed as follows:

$$C = Y_1 \tan\theta_2/\omega_0$$
$$\text{where} \quad Y_1 = 1/Z_1, \theta_2 = \pi/4 - \theta_1.$$

Looking from a different point of view, by replacing both θ_2 length transmission line components in (a) with lumped-element capacitors C as in (b), the two circuits are equivalent. The capacitor loaded UIR possesses the advantages of a small size and the capability of spurious response suppression. However, it is not always easy to apply the capacitor loaded UIR to frequency regions above 1 GHz, because the circuit loss of the lumped-element capacitor C increases dramatically as does the variance of resonance frequency, thus requiring frequency adjustment.

The loaded capacitance C can be replaced by an open-circuited transmission line. Furthermore, it is not always necessary to design the characteristic impedance of the transmission line at Z_1. An example is shown in (c) where the characteristic impedance is designed at $Z_2(= 1/Y_2)$. When $Y_2 \tan\theta_2' = Y_1\theta_2$, all three resonators will resonate at the same frequency. In this case, if $Z_2 < Z_1$, then, $\theta_2' < \theta_2$ thus the resonator length can be shortened. In addition, the elimination of the lumped-element capacitors enables the UIR to offer additional features including low-loss properties and a small amount of resonance frequency variance. Thus, by applying this design, the new model effectively overcomes the previously described weak points of the capacitor loaded UIR shown in (b).

This example illustrates that it is unnecessary to obtain uniform characteristic impedance of a transmission line resonator, and that resonator circuit performance can be realized by nonuniform impedance resonators (stepped impedance resonator: SIR). This is the basic structure and concept of the SIR which the authors address in this book [11]. Further discussion of the fundamental characteristics and practical applications of the SIR, including design examples, will be explained in detail below.

Although the structure of the SIR was widely understood, few practical applications to filters and oscillators were seen until the authors reported its availability. Although many reasons may be behind this lack of use, one main reason was that the technological importance of the SIR was not appreciated at the time. Another reason was merely design convenience, with many engineers deciding that the estimation of electrical performance was more practical with UIR than SIR. Currently, various demands for resonators in the RF/microwave area have made it necessary to select the most suitable resonator for each individual application. In addition, the use of CAD tools in the design process has become popular, enabling a remarkable reduction in calculation time of numerical analysis and optimization. These facts support the authors' belief that in the future, the application will determine the most suitable resonator to be selected among various types of transmission line resonators. Thus, it seems SIR, with its advanced structures, will be the first candidate among them.

The typical features of SIR, which will be discussed in further detail in the following chapters, are summarized as follows:

(1) A wide degree of freedom in structure and design.
(2) A wide range of applicable frequency through the use of various types of transmission lines (coaxial, stripline, microstrip, coplanar) and/or dielectric materials.
(3) Derivation of a generalized concept for transmission line resonators including UIR.
(4) Development of an expanded concept for nonuniform impedance resonators adopting advanced transmission line structures and composite materials.

Disregarding its simple structure, the SIR possesses numerous features and possibilities for practical application. The SIR can and will be applied not only to various filters but also to oscillators and mixers as a basic resonator in frequency bands from RF to millimeter wave. Such practical application will, in turn, help to verify the validity of SIR, thus solidifying its position as a familiar and conventional resonator device.

2. Basic Structure and Characteristics of SIR

Transmission-line resonators are most frequently used in frequency regions above the VHF band, yet, as described in the previous chapter, most applications employ structures with uniform characteristic impedance. Transmission-line resonators possessing a stepped-impedance structure (SIR) have long been known as an available resonator structure, and are often used to experimentally examine the effect of discontinuity in the impedance step of a transmission line [1]. However, this stepped impedance structure was hardly ever used for practical circuits, with the exception of impedance transformers.

Focusing on its capability of shortening resonator length without degradation of unloaded-Q, the authors verified its availability through theoretical analysis and actual fabrication of a duplexer circuit for mobile communications based on a $\lambda_g/4$-type co-axial SIR structure [2,3]. Ever since, numerous reports on SIR studies have been published, while research topics have expanded to various structures and types of SIR. Such activity has consequently brought remarkable progress to the development of various SIR applications.

In this part, the basic structure of $\lambda_g/4$, $\lambda_g/2$, and λ_g-type SIR are presented, followed by the introduction and definition of the impedance ratio R_Z, an important parameter for the analysis of the SIR. Next, basic properties such as resonance conditions, resonator length, spurious resonance frequencies and equivalent circuits are systematically discussed using R_Z. Such basic issues are presented to prepare the reader for discussions on actual and practical structures of SIR described in the following chapters.

2.1 Basic Structure of SIR

The SIR is a TEM or quasi-TEM mode resonator composed of more than two transmission lines with different characteristic impedance. Figure 2.1 shows typical examples of its structural variation in the case of the stripline configuration [4], where figures (a), (b), and (c) are, respectively, examples of $\lambda_g/4$, $\lambda_g/2$, and λ_g-type resonators. Alternative transmission-line structures other than the illustrated stripline configuration, such as coaxial and coplanar-line, are acceptable with the condition of a TEM/quasi-TEM mode resonance. Also, while the $\lambda_g/2$-type SIR shown in (b) employs an open-ended structure, short-circuited structures are also available.

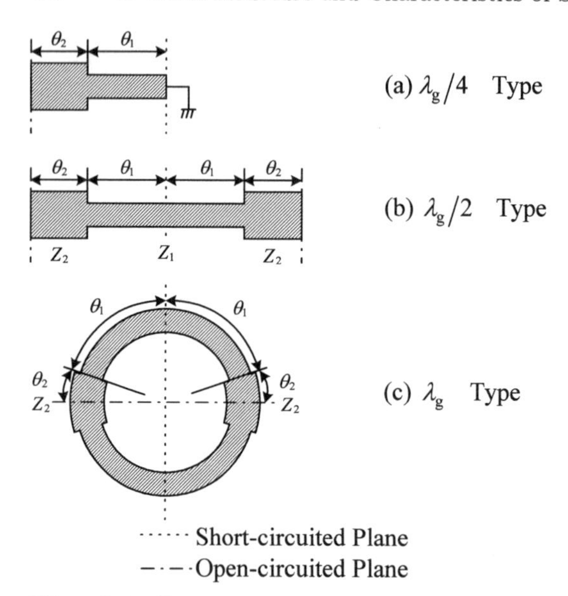

Fig. 2.1. Basic structures of SIR. (**a**) Quarter-wavelength type. (**b**) Half-wavelength type. (**c**) One-wavelength type

Characteristic impedance and corresponding electrical length of the transmission lines between the open- and short-circuited ends in Fig. 2.1 are defined as Z_1 and Z_2, θ_1 and θ_2, respectively (see figure).

The fundamental structural element common to all three types of SIR is a composite transmission line possessing both open- and short-circuited ends and a step junction in between. By defining this fundamental element, $\lambda_g/4$, $\lambda_g/2$, and λ_g-type SIR can, respectively, be looked upon as a combination of one, two, and four fundamental elements. An electrical parameter which characterizes the SIR is the ratio of the two transmission line impedances Z_1 and Z_2, and this we define by the following equation.

$$R_Z \equiv Z_2/Z_1 \qquad \text{impedance ratio} \tag{2.1}$$

It will be understood in the following section that R_Z is the most important parameter in characterizing properties of the SIR.

2.2 Resonance Conditions and Resonator Electrical Length

Figure 2.2 shows the fundamental element of SIR with an open-end, short-end and an impedance step. The input impedance and admittance are defined as Z_i and $Y_\mathrm{i}(= 1/Z_\mathrm{i})$, respectively. When ignoring the influences of step discontinuity and edge capacitance at the open end, Z_i can be expressed as follows [4]:

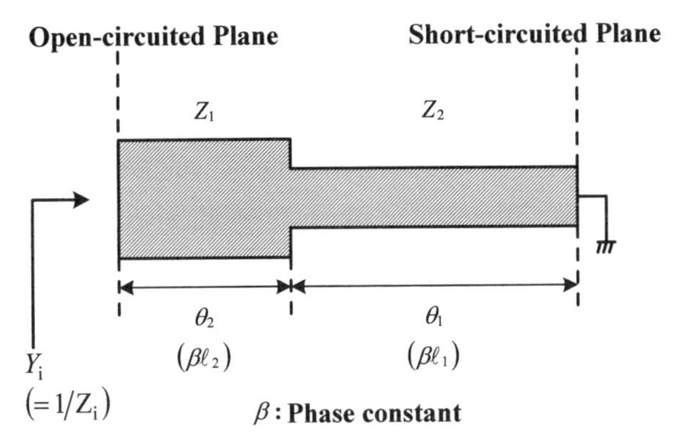

Fig. 2.2. Electrical parameters of elementary SIR

$$Z_{\mathrm{i}} = \mathrm{j}Z_2 \frac{Z_1 \tan\theta_1 + Z_2 \tan\theta_2}{Z_2 - Z_1 \tan\theta_1 \tan\theta_2}. \tag{2.2}$$

Let $Y_{\mathrm{i}} = 0$, then the parallel resonance condition can be obtained as follows:

$$Z_2 - Z_1 \tan\theta_1 \tan\theta_2 = 0,$$

thus

$$\tan\theta_1 \tan\theta_2 = Z_2/Z_1 = R_Z. \tag{2.3}$$

From the above equation, we understand that the resonance condition of SIR is determined by θ_1, θ_2 and impedance ratio R_Z. In the case of a conventional uniform impedance resonator (UIR), this condition is determined solely on transmission line length; however, for SIR, both the length and impedance ratio must be taken into account. This gives the SIR an extra degree of freedom in design as compared to the UIR.

The overall electrical length of the SIR, represented by θ_{TA}, is expressed as,

$$\begin{aligned}\theta_{\mathrm{TA}} &= \theta_1 + \theta_2 \\ &= \theta_1 + \tan^{-1}(R_Z/\tan\theta_1).\end{aligned} \tag{2.4}$$

Normalized resonator length is defined by the following equation with respect to the electrical length of a corresponding UIR measuring $\pi/2$.

$$\begin{aligned}L_{\mathrm{n}} &= \theta_{\mathrm{TA}}/(\pi/2), \\ &= 2\theta_{\mathrm{TA}}/\pi.\end{aligned} \tag{2.5}$$

Figure 2.3 shows the relationship between electrical length θ_1 and normalized resonator length L_{n} taking R_Z as parameter.

Overall electrical length of $\lambda_{\mathrm{g}}/2$- and λ_{g}-type SIR, defined, respectively, as θ_{TB} and θ_{TC}, can be expressed as,

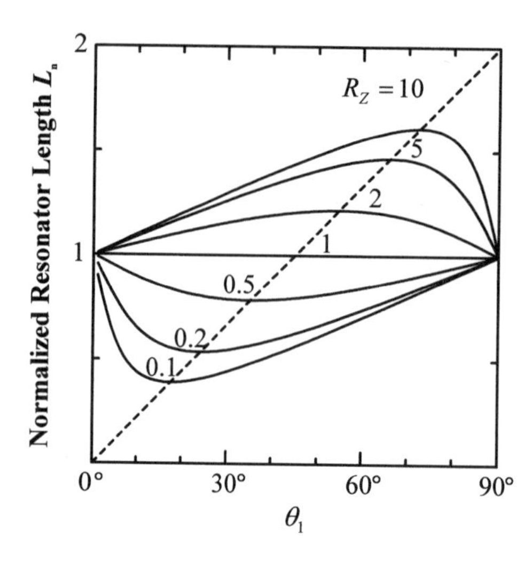

Fig. 2.3. Resonance condition of SIR

$$\theta_{\mathrm{TB}} = 2\theta_{\mathrm{TA}},$$
$$\theta_{\mathrm{TC}} = 4\theta_{\mathrm{TA}},$$

which can be normalized by corresponding UIR lengths of π and 2π, giving the following equations.

$$\theta_{\mathrm{TB}}/\pi = 2\theta_{\mathrm{TA}}/\pi = L_{\mathrm{n}}, \tag{2.6}$$
$$\theta_{\mathrm{TC}}/2\pi = 4\theta_{\mathrm{TA}}/2\pi = L_{\mathrm{n}}. \tag{2.7}$$

The above equations suggest that the resonance conditions of all three types of SIR can be expressed using the same equation. From Fig. 2.3 we understand that normalized resonator length L_{n} attains a maximum value when $R_Z \geq 1$ and a minimum value when $R_Z < 1$. Next, we examine the conditions which yield such maximum and minimum values. Substituting $\theta_2 = \theta_{\mathrm{TA}} - \theta_1$, (2.3) gives,

$$R_Z = \frac{\tan\theta_1(\tan\theta_{\mathrm{TA}} - \tan\theta_1)}{1 + \tan\theta_{\mathrm{TA}}\tan\theta_1}. \tag{2.8}$$

when $0 < R_Z < 1$ and $0 < \theta_{\mathrm{TA}} < \pi/2$,

$$\begin{aligned}\tan\theta_{\mathrm{TA}} &= \frac{1}{1 - R_Z}\left(\tan\theta_1 + \frac{R_Z}{\tan\theta_1}\right)\\ &= \frac{\sqrt{R_Z}}{1 - R_Z}\left(\frac{\tan\theta_1}{\sqrt{R_Z}} + \frac{\sqrt{R_Z}}{\tan\theta_1}\right)\\ &\geq \frac{2\sqrt{R_Z}}{1 - R_Z}.\end{aligned} \tag{2.9}$$

This expression is equated when $\tan\theta_1/\sqrt{R_Z} = \sqrt{R_Z}/\tan\theta_1$. This is equivalent to,

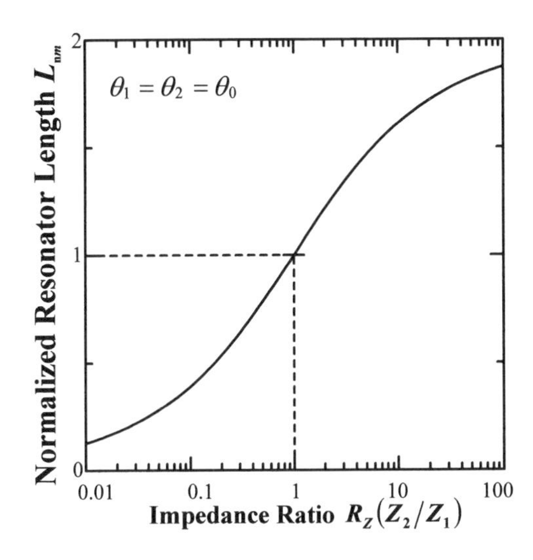

Fig. 2.4. Relationship between impedance ratio and normalized resonator length

$$\tan^2 \theta_1 = R_Z. \tag{2.10}$$

Thus, when

$$\theta_1 = \theta_2 = \tan^{-1} \sqrt{R_Z}. \tag{2.11}$$

θ_{TA} attains a minimum value of

$$(\theta_{\mathrm{TA}})_{\mathrm{min}} = \tan^{-1}\left(\frac{2\sqrt{R_Z}}{1 - R_Z}\right). \tag{2.12}$$

Similarly, if $R_Z > 1$ and $\pi/2 < \theta_T < \pi$, we obtain the following equation.

$$\tan\theta_{\mathrm{TA}} = -\frac{\sqrt{R_Z}}{R_Z - 1}\left(\frac{\tan\theta_1}{\sqrt{R_Z}} + \frac{\sqrt{R_Z}}{\tan\theta_1}\right). \tag{2.13}$$

Due to $0 < \theta_1 < \pi/2, \theta_{\mathrm{TA}}$ attains a maximum value of

$$(\theta_{\mathrm{TA}})_{\mathrm{max}} = \tan^{-1}\left(\frac{2\sqrt{R_Z}}{1 - R_Z}\right) \qquad \text{when} \qquad \theta_1 = \theta_2 = \tan^{-1}\sqrt{R_Z}.$$

These calculations reveal that $\theta_1 = \theta_2$ expresses a special condition which yields SIR maximum or minimum length. The following discussions are thus based mainly on this condition. Figure 2.4 shows the relationship between impedance ratio R_Z and normalized resonator length L_{n0} when $\theta_1 = \theta_2 \equiv \theta_0$. Here, L_{n0} can be expressed as follows:

$$L_{\mathrm{n0}} = 2\theta_{\mathrm{TA}}/\pi = 4\theta_0/\pi = 4\left(\tan^{-1}\sqrt{R_Z}\right)/\pi. \tag{2.14}$$

Figure 2.4 illustrates the fact that resonator length can be infinitely shortened by applying a smaller R_Z value, while maximum resonator length is limited to twice that of the corresponding UIR.

2.3 Spurious Resonance Frequencies

A distinct feature of the SIR is that the resonator length and corresponding spurious resonance frequencies can be adjusted by changing the impedance ratio R_Z [4]. In the following discussion, the fundamental resonance frequency is represented as f_0, while the lowest spurious frequencies of $\lambda_g/4$, $\lambda_g/2$, and λ_g-type SIR are represented, respectively, as f_{SA}, f_{SB}, and f_{SC}. For understanding the general tendency of spurious responses, we consider the TEM mode as the dominant resonance mode, and neglect the effect of the step junction in the transmission line of the resonator. We also assume $\theta_1 = \theta_2 = \theta_0$ as the structure of SIR.

Resonator electrical lengths corresponding to spurious frequencies f_{SA}, f_{SB}, and f_{SC} are expressed as θ_{SA}, θ_{SB}, and θ_{SC}, respectively.

From (2.3), the following equation is obtained for f_{SA}.

$$\tan\theta_{SA} = \tan(\pi - \theta_0) = -\tan^{-1}\sqrt{R_Z}. \tag{2.15}$$

The resonance conditions for $\lambda_g/2$ and λ_g-type SIR can be derived from the following equation.

$$(R_Z \tan\theta_1 + \tan\theta_2)(R_Z - \tan\theta_1 \cdot \tan\theta_2) = 0.$$

Considering $\theta_1 = \theta_2 = \theta$ gives

$$\tan\theta \cdot (R_Z + 1)(R - \tan^2\theta) = 0.$$

From the above equation we obtain the following solutions.

$$\begin{aligned} \theta_0 &= \tan^{-1}\sqrt{R_Z}, \\ \theta_{SB} &= \theta_{SC} = \pi/2. \end{aligned} \tag{2.16}$$

Consequently, spurious resonance frequencies are obtained as follows:

$$\frac{f_{SA}}{f_0} = \frac{\theta_{SA}}{\theta_0} = \frac{\pi - \theta_0}{\theta_0} = \frac{\pi}{\tan^{-1}\sqrt{R_Z}} - 1, \tag{2.17}$$

$$\frac{f_{SB}}{f_0} = \frac{f_{SC}}{f_0} = \frac{\theta_{SB}}{\theta_0} = \frac{\pi}{2\tan^{-1}\sqrt{R_Z}}. \tag{2.18}$$

Figure 2.5 illustrates the above relationship. Generally speaking, the fundamental and spurious frequencies should be as far apart as possible. Thus, the figure suggests that a $\lambda_g/4$-type SIR accompanying a small R_Z value should be chosen for optimal design, which simultaneously minimizes resonator length.

2.4 Derivation of an Equivalent Circuit of SIR

High-frequency RF circuits often require the use of distributed-element components for practical fabrication. Even so, analysis using equivalent lumped-element descriptions, rather than the direct analysis of distributed circuits,

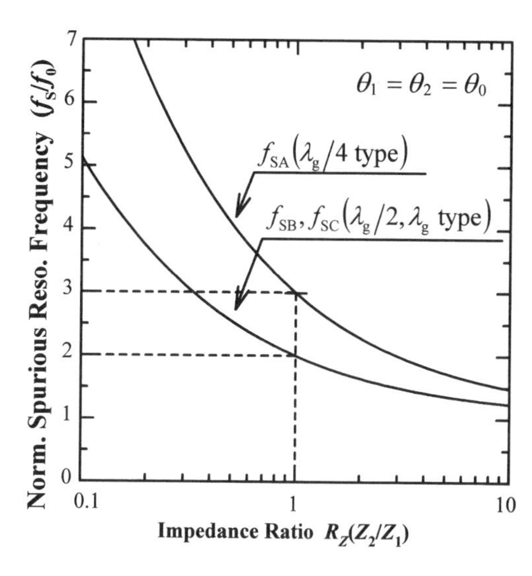

Fig. 2.5. Relationship between impedance ratio and normalized spurious resonance frequencies

turns out to be far more convenient and predictable. Furthermore, basic design theories of bandpass filters are based on lumped-element circuits, thus making it possible to apply these results to the design of such distributed-element resonators. In the resonating state, a distributed-element resonator such as SIR can be approximately expressed using equivalent lumped-elements L, C, R [4], which are derived from the slope parameters of the resonator. The susceptance slope parameter b_s can be obtained from its definition as follows:

$$b_s \equiv \frac{\omega_0}{2} \cdot \frac{dB_s}{d\omega}\Big|_{\omega=\omega_0}, \tag{2.19}$$

where w_0 is the angular resonance frequency and $B_s(w)$ the susceptance of the resonator. Let the susceptance of a $\lambda_g/4$-type SIR be B_{SA} and the corresponding slope parameter be b_{SA}, the slope parameter is derived from (2.2) and (2.19) as follows:

$$\begin{aligned} B_{SA} &= \mathrm{Im}[1/Z_i] \\ &= Y_2 \cdot \frac{\tan\theta_1 \cdot \tan\theta_2 - R_Z}{\tan\theta_1 + R_Z \tan\theta_2}, \end{aligned} \tag{2.20}$$

$$\begin{aligned} b_{SA} &= \frac{\omega_0}{2} \cdot \frac{dB}{d\omega}\Big|_{\omega=\omega_0} \\ &= \frac{\theta_0}{2} \cdot \frac{dB}{d\theta}\Big|_{\theta=\theta_0} \\ &= \frac{\theta_{01} Y_2}{2} \cdot \left[\frac{R_Z}{(1 - R_Z{}^2)\sin^2\theta_{01} + R_Z{}^2} + \frac{\ell_1}{\ell_2}\right], \end{aligned} \tag{2.21}$$

where θ_{01} expresses the value of θ_1 at resonance state and ℓ_1 and ℓ_2 show resonator physical length. In the case of $\theta_{01} = \theta_{02} = \theta_0$, i.e. $\ell_1 = \ell_2$, considering

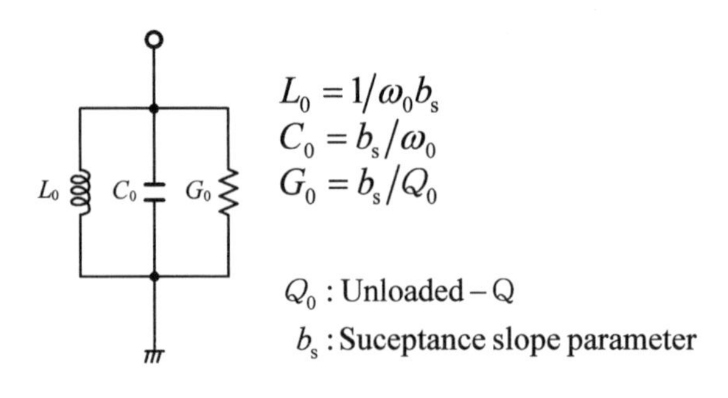

Fig. 2.6. An equivalent circuit of SIR at resonance

$\tan^2\theta_0 = R_Z$ and $\sin^2\theta_0 = R_Z/(1+R_Z)$, then

$$\begin{aligned} b_{SA0} &= \frac{\theta_0 Y_2}{2}\cdot\left[\frac{R_Z}{R_Z(1-R_Z)+R_Z{}^2}+1\right] \\ &= \theta_0 Y_2. \\ &= Y_2\tan^{-1}\sqrt{R_Z}. \end{aligned} \tag{2.22}$$

Similarly, the susceptance slope parameter of $\lambda_g/2$ and λ_g-type SIRs, represented as b_{SB0} and b_{SC0} when $\theta_1 = \theta_2 = \theta_0$, can be obtained as follows:

$$b_{SB0} = 2\theta_0 Y_2 = 2Y_2\tan^{-1}\sqrt{R_Z}, \tag{2.23}$$

$$b_{SC0} = 4\theta_0 Y_2 = 4Y_2\tan^{-1}\sqrt{R_Z}. \tag{2.24}$$

The relationship between the susceptance slope parameter and L_0, C_0 and G_0 of a lumped-element resonator are given as follows:

$$L_0 = 1/\omega_0 b_s,\ C_0 = b_s/\omega_0,\ G_0 = b_s/Q_0, \tag{2.25}$$

where Q_0 represents the unloaded-Q of SIR.

Therefore, the SIR equivalent circuit near resonance frequency can be expressed in the form illustrated in Fig. 2.6. The unloaded-Q is determined by physical dimension and the materials composing the resonator, thus it must be calculated using another method, which will be discussed in the following chapters.

3. Quarter-Wavelength-Type SIR

The basic characteristics and practical applications of a $\lambda_g/4$-type SIR are described in this chapter. From an application-oriented viewpoint, the $\lambda_g/4$-type SIR is the most attractive among various types of SIR, and numerous R&D results have been reported on this type. The main reason for such interest is that the $\lambda_g/4$-type SIR has an essential availability for resonator miniaturization [1]. As discussed in Chap. 2, one of the main features of the $\lambda_g/4$-type SIR is the capability of reducing resonator size, while another is the capability of controlling spurious frequencies by design. These two properties make the $\lambda_g/4$-type SIR an extremely suitable resonator element for mobile communication filters. A coaxial structure is most frequently adopted for practical application of $\lambda_g/4$-type SIR to filters. Coaxial type resonators have the following advantages.

1) High Q values compared with lumped-element or helical resonators;
2) Simple structure and manufacturing convenience;
3) A wide variation of applicable coupling methods;
4) High power handling capabilities.

These attractive features have resulted in extensive research focused on filters and duplexers used in wireless communication, especially in mobile communication-oriented applications that require miniaturization. Bandpass filters with wide stop-band characteristics can easily be realized by utilizing its capability of controlling spurious responses.

In this part, features of the $\lambda_g/4$-type SIR are summarized, followed by a discussion on problems encountered in designing coaxial type SIR. Resonance conditions considering the effects of step discontinuity and open-end are derived, and the unloaded-Q value, indicating the "figure of merit" of the resonator, is also discussed.

Next, after organizing the filter design method using SIR, practical design examples using air-cavity type and dielectric-type coaxial SIR are introduced, and finally the stripline SIR is examined.

3.1 Analysis of $\lambda_g/4$-Type Coaxial SIR

3.1.1 Impedance Ratio R_Z

Figure 3.1 shows the fundamental structure of a coaxial type SIR discussed hereafter. Figure 3.1a illustrates a structure with a constant outer-conductor diameter and a step junction in the inner-conductor, while Fig. 3.1b shows a contrary structure possessing a constant inner-conductor diameter and a step junction in the outer-conductor. The outer diameter of the coaxial inner-conductor and the inner diameter of the outer-conductor are expressed as $2a$ and $2b$, respectively. When the relative dielectric constant of the dielectric filling the space between inner and outer conductors is defined as ε_r, the characteristic impedance of the coaxial transmission line can be expressed as follows:

$$Z = \frac{60}{\sqrt{\varepsilon_r}} \ln\left(\frac{b}{a}\right) . \tag{3.1}$$

Therefore, the two transmission line impedances for structure (a) can be obtained as follows:

$$Z_1 = \frac{60}{\sqrt{\varepsilon_r}} \ln\left(\frac{b}{a_1}\right) ,$$

$$Z_2 = \frac{60}{\sqrt{\varepsilon_r}} \ln\left(\frac{b}{a_2}\right) .$$

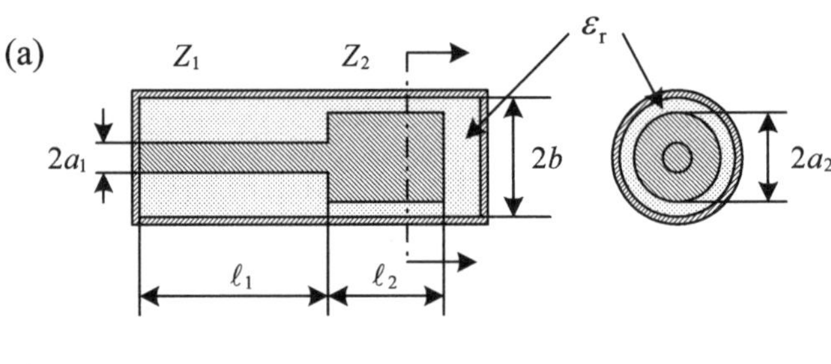

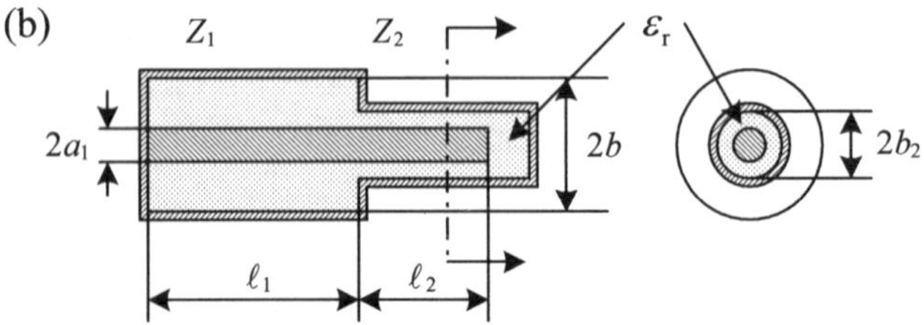

Fig. 3.1. Fundamental structures of coaxial quarter-wavelength type SIR. (**a**) A structure with a constant outer-conductor diameter. (**b**) A contrary structure with a constant inner-conductor diameter

Thus, the impedance ratio R_Z is derived as:

$$R_Z = Z_2/Z_1 = \ln(b/a_2)/\ln(b/a_1). \tag{3.2}$$

Similarly, R_Z for structure (b) is given as follows:

$$R_Z = Z_2/Z_1 = \ln(b_2/a_1)/\ln(b/a_1). \tag{3.3}$$

3.1.2 Effects of Discontinuity

As discussed in Chap. 2, discontinuities in the transmission-line of the SIR must be taken into consideration when obtaining an accurate expression for SIR resonance frequency. Figure 3.2 illustrates two such discontinuities; one is the open-end capacitance C_f, and the other is the discontinuity capacitance C_d due to a step junction in the transmission line. As shown in the figure, the open-end fringing capacitance C_f of a coaxial transmission line can be expressed as follows [2,3]:

$$C_f = C_p + C_{fe} + C_{fs} \tag{3.4}$$

where $C_p = \pi a^2 \varepsilon_0 / d$: primary capacitance
C_{fe} : end-wall fringing capacitance
C_{fs} : side-wall fringing capacitance.

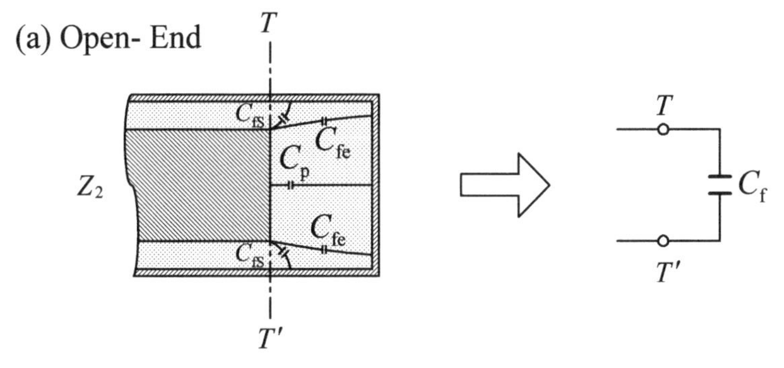

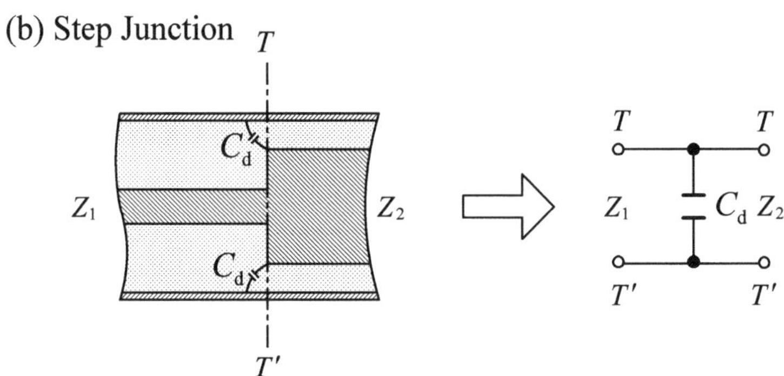

Fig. 3.2. Discontinuities in SIR. (**a**) Open-end fringing effect. (**b**) Step-junction effect

In the case of a coaxial SIR, obtaining resonator miniaturization requires the transmission line impedance Z_2 to be designed at as small a value as possible. Therefore, in practical design, the distance from the center conductor to the outer conductor, $2(b-a_2)$, tends to be much narrower than the gap d between the open-end of the center conductor and the side-wall. This enables C_f to be approximated as follows:

$$C_\mathrm{f} \cong C_\mathrm{fs}.$$

The fringing capacitance C_fs of a coaxial transmission line can be derived from an equivalent equation for parallel plate capacitors by applying a structural transformation [4]:

$$C_\mathrm{fs} = \frac{b\varepsilon_0\varepsilon_r}{t\ln(b/a_2)}\left[2t\ln(1+t) - (t-1)\ln(t^2-1)\right] \tag{3.5}$$

where $t = 1/(1 - a_2/b)$

ε_0 : dielectric constant (permittivity) in free space,

ε_r : relative dielectric constant.

Converting this open-end capacitance into equivalent transmission line length proves to be convenient for a practical design. This can be calculated from the following relationship:

$$\mathrm{j}\omega C_\mathrm{f} \cong \mathrm{j}Y_2\tan(\beta\Delta\ell). \tag{3.6}$$

Considering $\beta\Delta\ell$ a minimal value,

$$\Delta\ell \cong \omega C_\mathrm{f}/Y_2\beta,$$

$$\text{where}\quad Y_2 = \frac{2\pi}{\ln(b/a_2)}\cdot\sqrt{\frac{\varepsilon_0}{\mu_0}},\qquad \beta = \omega\sqrt{\varepsilon_0\mu_0}$$

μ_0 : permittivity in free space.

Thus, when $C_\mathrm{f} \cong C_\mathrm{fs}$, $\Delta\ell$ can be expressed as follows:

$$\begin{aligned}\Delta\ell &\cong \frac{C_\mathrm{fs}}{2\pi\varepsilon_0}\ln(b/a_2)\\ &= \frac{b}{2\pi t}\left[2t\ln(1+t) - (t-1)\ln(t^2-1)\right].\end{aligned} \tag{3.7}$$

Thus, the fringing effects can be translated into an increase in line length instead of directly considering fringing capacitance. For example, in the case of a SIR with $2b = 15\,\mathrm{mm}$ and $2a_2 = 12\,\mathrm{mm}$, the equivalent line length is obtained as $\Delta\ell \approx 0.17b = 1.3\,\mathrm{mm}$.

The next discussion focuses on the effect caused by a step junction in the transmission line of a resonator. This issue has been thoroughly investigated, and such research results indicate that the discontinuity at a step junction can be expressed as an equivalent capacitance C_d. Such results have been summarized and prepared as design charts [5], enabling one to easily obtain the C_d value according to a given step-junction structure and dielectric material. For example, in the case of an air-cavity SIR with dimensional features of $2b = 15\,\text{mm}$, $2a_1 = 4\,\text{mm}$, and $2a_2 = 12\,\text{mm}$, the approximate value of C_d is given as 0.28 pF. A more direct approach in obtaining the C_d value is through numerical analysis of the electromagnetic field distribution at the step junction, however, this requires time-consuming calculations. Considering the above-mentioned discontinuities, namely the open-end and step junction, we derive a more accurate expression for the resonance condition of a SIR. As previously discussed, the open-end fringing capacitance can be interpreted as an equivalent line length $\Delta\ell$, thus enabling θ_2 to be expressed as follows:

$$\theta_2' = \beta(\ell_2 + \Delta\ell). \tag{3.8}$$

The resonance condition of a SIR possessing a discontinuity capacitance at the step junction of the transmission line is given by;

$$\frac{\tan\theta_1 \tan\theta_2}{1 - \alpha_0 \tan\theta_1} = \frac{Z_1}{Z_2} = R_Z, \tag{3.9}$$

$$\text{where} \quad \alpha_0 = \omega_0 C_d Z_1$$

$$\omega_0 : \text{angular resonance frequency.}$$

For example, in the case of an air-cavity type SIR designed at 1000 MHz with dimensional features of $2b = 15\,\text{mm}$, $2a_1 = 4\,\text{mm}$, and $2a_2 = 12\,\text{mm}$, R_Z is given as follows:

$$\begin{aligned} Z_1 &= 60\ln(b/a_1) = 79.3\,\Omega, \\ Z_2 &= 60\ln(b/a_2) = 13.4\,\Omega, \\ R_Z &= Z_2/Z_1 = 0.169. \end{aligned}$$

From the previous estimations, $\Delta\ell$ and C_d are respectively 13 mm and 0.28 pF, thus giving $\alpha_0 = 0.140$. Let $\theta_1 = \theta_2'$, then ℓ_1 can be calculated from (3.9) as 17.9 mm, and thus $\ell_2 = \ell_1 - \Delta\ell = 16.9\,\text{mm}$, while $\ell_T = \ell_1 + \ell_2 = 34.8\,\text{mm}$. This is compared with the total resonator length neglecting the influence of discontinuity, which is obtained from Fig. 2.3 as $\ell_T = 37.2\,\text{mm}$. These calculations illustrate that in the actual design of this specific resonator, resonator length must be designed 6% shorter in order to cancel the effects due to discontinuity. In addition, (3.7) suggests that $\Delta\ell$ is determined by structural parameters and possesses no frequency dependency. C_d is also determined by structure and is independent of frequency itself, however, it has a direct relation with α_0, thus effecting the resonance frequency of the SIR. The higher the resonance frequency, the greater its influence. We understand from this discussion that high frequency applications demand rigorous estimation of the effects due to discontinuities within the resonator structure.

3.1.3 Unloaded-Q of Coaxial SIR

Unloaded-Q is frequently used in discussing the "figure of merit" of a resonator. In broad terms, the unloaded-Q is a significant parameter applied in circuit design for the evaluation of electrical performance, and is generally obtained by numerical analysis using high-speed computers. In the case of a coaxial resonator, closed form equations for unloaded-Q can be derived by analytical methods, thus enabling more effective and foresighted evaluation of Q values when compared with numerical analysis. In this chapter we derive the unloaded-Q of a SIR from the geeral definition of Q. Unloaded-Q(Q_0) for a resonant circuit is defined as follows:

$$Q_0 = \omega_0 \cdot \frac{W_\mathrm{S}}{P_\mathrm{L}}, \tag{3.10}$$

where W_S : maximum energy stored in the resonant circuit,
P_L : average power loss in the resonant circuit,
ω_0 : angular resonance frequency.

When $\boldsymbol{E}$ and $\boldsymbol{H}$ respectively, represent the electric field and magnetic field distributed in the resonator, W_s can be expressed as:

$$W_\mathrm{s} = \frac{\mu}{2} \int_V |H|^2 \,\mathrm{d}V = \frac{\varepsilon}{2} \int_V |E|^2 \,\mathrm{d}V \tag{3.11}$$

where $\mu = \mu_0 \mu_\mathrm{r}$: permeability of the material,
$\varepsilon = \varepsilon_0 \varepsilon_\mathrm{r}$: permittivity (dielectric constant) of the material.

Assuming that the dielectric material of the resonator in this discussion employs a low loss characteristic, we substitute $\mu = \mu_0$ and, furthermore, neglect magnetic loss. Consequently, P_L is expressed as follows:

$$P_\mathrm{L} = P_\mathrm{LC} + P_\mathrm{LD} \tag{3.12}$$

where P_LC : side-wall surface resistivity loss of the resonator,
P_LD : dielectric loss of the material.

By expressing dielectric constant using real and imaginary components as $\varepsilon = \varepsilon' + \mathrm{j}\varepsilon''$,

$$P_\mathrm{LD} = \frac{\omega_0 \varepsilon''}{2} \int_V |E|^2 \,\mathrm{d}V. \tag{3.13}$$

Considering $\varepsilon'/\varepsilon'' = \tan \delta_\mathrm{d}$ (loss tangent), the above equation becomes

$$P_\mathrm{LD} = \frac{\omega_0 \varepsilon' \tan \delta_\mathrm{d}}{2} \int_V |E|^2 \,\mathrm{d}V.$$

Thus, unloaded-Q is derived as follows:

$$\begin{aligned}\frac{1}{Q_0} &= \frac{P_L}{\omega_0 W_S} \\ &= \frac{P_{LC}}{\omega_0 W_S} + \frac{P_{LD}}{\omega_0 W_S} \\ &= \frac{R_S}{2}\int_S |H_t|^2\,dS \Big/ \frac{\mu_0\omega_0}{2}\int_V |H|^2\,dV \\ &\quad + \frac{\omega_0\varepsilon' \tan\delta_d}{2}\int_V |E|^2\,dV \Big/ \frac{\varepsilon\omega_0}{2}\int_V |E|^2\,dV \end{aligned} \tag{3.14}$$

where R_S : surface resistivity of the conductor material,
H_t : tangential component of H at the conductor surface.

The first component in (3.14) is expressed using Q_C as

$$\frac{1}{Q_C} = \frac{R_S}{\mu_0\omega_0}\int_S |H_t|^2\,dS \Big/ \int_V |H|^2\,dV. \tag{3.15}$$

The above Q_C is determined by the side-wall conductor loss of the resonator. Furthermore, considering a small $\tan\delta_d$, $\varepsilon = \varepsilon' + j\varepsilon'' \cong \varepsilon'$ is assumed in order to simplify the second component representing the effect of the electric field. Consequently, (3.14) results in the following equation:

$$\begin{aligned}\frac{1}{Q_0} &= \frac{1}{Q_C} + \frac{1}{Q_d} \\ &= \frac{1}{Q_C} + \tan\delta_d,\end{aligned} \tag{3.16}$$

where $\tan\delta_d = 1/Q_d$.

Thus Q_0 can be obtained from Q_C.

The dominant resonant mode of a coaxial resonator being TEM mode, we approximate the magnetic field by concentrating only on its radial component H_ϕ. Expressing the current in the center conductor as I, H_ϕ is expressed as follows:

$$H_\phi = \frac{I}{2\pi r} \quad (a \le r \le b).$$

Considering the current distribution in the SIR, Q_C can be obtained by integrating H_ϕ according to (3.15). This is illustrated for the SIR structure shown in Fig. 3.1a. The SIR is divided into two regions of differing transmission line impedance Z_1 and Z_2, respectively represented as region (I) and region (II), and maximum stored energy W_S within each region is expressed as W_{S1} and W_{S2}. Assuming I_0 the current at the short-circuited end of the SIR, current $I(x)$ in region (I) is expressed as

$$I(x) = I_0\cos\beta x \quad (x = 0 : \text{short-circuited point}).$$

Let inductance per unit length be L_1, then the stored energy between x and $x + dx$ can be expressed as:

$$\frac{1}{2}L_1 \mathrm{d}x\,|I(x)|^2 = \frac{L_1 I_0^2}{2}\cos^2 \beta x \mathrm{d}x$$
$$\text{where} \quad L_1 = \frac{\mu_0}{2\pi}\ln(b/a_1).$$

Thus, integrating the above equation gives

$$\begin{aligned} W_{S1} &= \frac{L_1 I_0^2}{2}\int_0^{\ell_1}\cos^2 \beta x \mathrm{d}x \\ &= \frac{L_1 I_0^2}{8\beta}(\sin 2\theta_{01} + 2\theta_{01}). \end{aligned} \tag{3.17}$$

Focusing next on W_{S2} in region (II), the following equation is obtained when considering current continuity at the step junction:

$$I(x) = \frac{\cos\theta_{01}}{\sin\theta_{02}} I_0 \sin \beta x.$$

Again, by integrating the above equation,

$$W_{S2} = \frac{L_2 I_0^2}{8\beta}\frac{\cos^2\theta_{01}}{\sin^2\theta_{02}}(2\theta_{02} - \sin 2\theta_{02}), \tag{3.18}$$
$$\text{where} \quad L_2 = \frac{\mu_0}{2\pi}\ln(b/a_2).$$

Thus, total stored energy W_S is derived from (3.17) and (3.18) as follows:

$$\begin{aligned} W_S &= W_{S1} + W_{S2} \\ &= \frac{I_0^2}{8\beta}\left[L_1(2\theta_{01} + \sin 2\theta_{01}) + L_2\frac{\cos^2\theta_{01}(2\theta_{02} - \sin 2\theta_{02})}{\sin^2\theta_{02}}\right]. \end{aligned} \tag{3.19}$$

Discussions further proceed onto power losses within the SIR. Power losses are generated at the surface of the conductor at regions (I) and (II), the short-circuited end, and the step junction of the transmission, defined P_{LC1}, P_{LC2}, P_{LSC}, and P_{LSJ}, respectively. The resistance per unit length of the transmission line at regions (I) and (II), represented as R_1 and R_2, are expressed as follows:

$$R_1 = \frac{R_S}{2\pi}\left(\frac{1}{a_1} + \frac{1}{b}\right) = \frac{R_S}{2\pi b}\left(1 + \frac{b}{a_1}\right),$$
$$R_2 = \frac{R_S}{2\pi}\left(\frac{1}{a_2} + \frac{1}{b}\right) = \frac{R_S}{2\pi b}\left(1 + \frac{b}{a_2}\right),$$

where $R_S = 1/\sigma\delta$: surface resistivity,
σ : conductivity of the material,
$\delta = 1/\sqrt{\omega_0\mu_0\sigma/2}$: skin depth.

Thus, P_{LC1} and P_{LC2} are obtained from the following equations.

$$P_{\mathrm{LC1}} = \frac{R_1}{2}\int_0^{\ell_1} |I(x)|^2 \, \mathrm{d}x = \frac{R_1 I_0^2}{8\beta}(2\theta_{01} + \sin 2\theta_{01}), \tag{3.20}$$

$$P_{\mathrm{LC2}} = \frac{R_2}{2}\int_{\ell_1}^{\ell_1+\ell_2} |I(x)|^2 \, \mathrm{d}x = \frac{R_2 I_0^2}{8\beta}\frac{\cos^2\theta_{01}}{\sin^2\theta_{02}}(2\theta_{02} - \sin 2\theta_{02}) \cdot \tag{3.21}$$

Let the resistance at the short-circuited plane be R_{01}, then

$$R_{01} = \int_{a_1}^{b} \frac{R_S}{2\pi r}\mathrm{d}r = \frac{R_S}{2\pi}\ln(b/a_1).$$

The current at the short-circuited end being I_0,

$$P_{\mathrm{LCS}} = \frac{1}{2}R_{01}I_0^2 = \frac{R_S I_0^2}{4\pi}\ln(b/a_1). \tag{3.22}$$

Similarly, the resistance at the step junction, represented by R_{02}, becomes

$$R_{02} = \int_{a_1}^{a_2} \frac{R_S}{2\pi r}\mathrm{d}r = \frac{R_S}{2\pi}\ln(a_2/a_1) = \frac{R_S}{2\pi}\left[\ln(b/a_1) - \ln(b/a_2)\right].$$

The current at the step junction, being $I_0 \cos\theta_{01}$, P_{LCJ} is obtained as follows:

$$P_{\mathrm{LCJ}} = \frac{1}{2}R_{01}I_0^2\cos^2\theta_{01} = \frac{R_S}{4\pi}I_0^2\cos^2\theta_{01}\left[\ln(b/a_1) - \ln(b/a_2)\right] \cdot \tag{3.23}$$

Therefore, total loss of the SIR results in the following equation.

$$\begin{aligned} P_{\mathrm{LC}} &= P_{\mathrm{LC1}} + P_{\mathrm{LC2}} + P_{\mathrm{LCS}} + P_{\mathrm{LCJ}} \\ &= \frac{I_0^2}{8\beta}(A_1R_1 + A_2B_2R_2) + \frac{I_0^2}{2}(R_{01} + B_1), \end{aligned} \tag{3.24}$$

where $A_1 = 2\theta_{01} + \sin 2\theta_{01}$, $A_2 = 2\theta_{02} - \sin\theta_{02}$,
$B_1 = \cos^2\theta_{01}$, $B_2 = \cos^2\theta_{01}/\sin^2\theta_{02}$.

Thus Q_{C} is obtained from (3.10), (3.19), and (3..24), as follows:

$$Q_{\mathrm{C}} = \omega_0\frac{W_{\mathrm{S}}}{P_{\mathrm{LC}}} = \frac{2b}{\delta}\cdot\frac{C_1}{D_1} \tag{3.25}$$

where
$$\begin{aligned} C_1 &= A_1 + A_2B_2\ln(b/a_2), \\ D_1 &= A_1(1 + b/a_1) + A_2B_2(1 + b/a_2) \\ &\quad + 4\beta b\left\{\ln(b/a_1) + B_1\left[\ln(b/a_1) - \ln(b/a_2)\right]\right\}. \end{aligned}$$

When dielectric loss is considered sufficiently small, as in the case of an air-cavity type resonator, (3.16) can be approximated as $Q_{\mathrm{C}} \cong Q_0$.

Furthermore, (3.25) is simplified to reveal an approximate relationship between Q_{C} and resonator structure. We start this discussion by considering

$Q_C(= Q_{CU})$ of a $\lambda_g/4$ type UIR. In this case, it is apparent that line length $\theta_{01} = \theta_{02} = \pi/4$ and center conductor radius $a_1 = a_2 = a$, thus

$$A_1 = A_2 = \pi/2, \quad B_1 = 1/2, \quad B_2 = 1.$$

Consequently, C_1 and D_1 are expressed as

$$C_1 = \pi \ln(b/a),$$
$$D_1 = \pi(1 + b/a) + 4\beta b \ln(b/a) = \pi(1 + b/a) + (2\pi b/\lambda_0) \ln(b/a).$$

Q_{CU} is obtained accordingly as

$$Q_{CU} = \frac{2b}{\delta} \cdot \frac{\pi \ln(b/a)}{\pi(1 + b/a) + (8b/\lambda_0) \ln(b/a)}. \tag{3.26}$$

For simplicity we ignore the final term of the denominator which represents the loss at the short-circuited end, consequently obtaining

$$Q_{CU} \approx \frac{2b}{\delta} \cdot \frac{\ln(b/a)}{1 + b/a}. \tag{3.27}$$

Q_{CU}, expressed as a function of b/a, represents a generalized Q_C of a coaxial UIR with two open-circuited ends. It is well known that Q_{CU} attains a maximum value at $b/a = 3.6$. This means that the unloaded-Q of a coaxial UIR attains a maximum value when the characteristic impedance takes the following value:

$$Z_1 = (60/\sqrt{\varepsilon_r}) \ln(b/a) = 77/\sqrt{\varepsilon_r} \qquad \Omega.$$

Representing Q_{CU} at this maximum condition as Q_{CM}, this calculates to

$$Q_{CM} = 0.557b/\delta.$$

When applying copper as the conductor material, the above equation results in,

$$Q_{CM} = 2670b\sqrt{f_0},$$

where b: in cm,

f_0: resonance frequency in GHz.

Next, we examine the approximate Q value of SIR. Here, effects due to losses generated at a short-circuited end and a step junction are neglected as in the case of UIR. Let $\theta_{01} = \theta_{02}$, and considering the following relations,

$$R_Z = \tan^2\theta_{01} = \tan^2\theta_{02}, \quad \sin^2\theta_{01} = \sin^2\theta_{02} = R_Z/(1 + R_Z),$$
$$\cos^2\theta_{01} = \cos^2\theta_{02} = 1/(1 + R_Z), \quad \ln(b/a_1) = Z_1\sqrt{\varepsilon_r}/60,$$
$$\text{and} \quad \ln(b/a_2) = Z_2\sqrt{\varepsilon_r}/60,$$

we obtain the following equation by arranging (3.25),

$$Q_{CS} = \frac{b}{30\delta} \cdot \frac{C_2}{D_2}, \tag{3.28}$$

where $C_2 = 4Z_1\theta_{01} = 4Z_1 \tan^{-1}\sqrt{R_Z}$,

$$D_2 = \left(\frac{2\sqrt{R_Z}}{1+R_Z} + 2\theta_{01}\right)\left[1 + \exp\left(\frac{Z_1\sqrt{\varepsilon_r}}{\varepsilon_0}\right)\right] + \frac{1}{R_Z}\left(2\theta_{01} - \frac{2\sqrt{R_Z}}{1+R_Z}\right)\left[1 + \exp\left(\frac{R_Z Z_1\sqrt{\varepsilon_r}}{\varepsilon_0}\right)\right].$$

Thus Q_{CN}, the Q value normalized by Q_{CM}, can be expressed as

$$Q_{CN} = Q_{CS}/Q_{CM}. \tag{3.29}$$

Figure 3.3 illustrates the relationship between Q_{CN} and Z_1 taking R_Z as parameter. $R_Z = 1$ in the figure indicates the calculated curve for a coaxial UIR, yielding a maximum value at $Z_1 = 77\,\Omega$ as previously described. The Q value for SIR also possesses a maximum value, where Z_1 giving maximum Q value shows a slight upward shift as R_Z decreases. Moreover, the maximum Q_{CN} value itself decreases in accordance with R_Z. This implies that the Q value will degrade when resonator length shortens, as shown in Fig. 3.3. However, research results have theoretically and experimentally proven that the degree of Q value degradation due to miniaturization is significantly smaller for the SIR when compared to conventional resonators such as capacitor-loaded UIR.

As previously mentioned, the results illustrated above lack accuracy because influences of losses generated at the short-circuited ends and the step junction are neglected, however, a general relationship between Q_C and SIR structure can be conceptually understood. Moreover, the example shown in Fig 3.3 is merely an analytical result assuming $\ell_1 = \ell_2$; it does not indicate the maximum Q value under the condition of a limited total SIR length. Other reports claim an improved Q value when $\ell_1 > \ell_2$. Such structural conditions yielding a maximum Q value will be discussed further in Chap. 6.

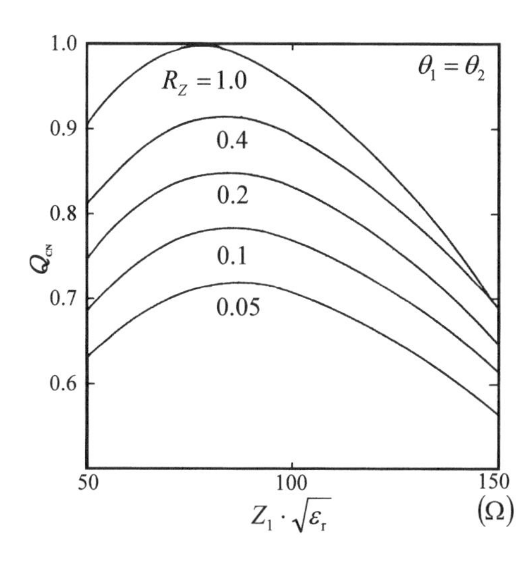

Fig. 3.3. Relationship between normalized-Q and line impedance Z_i

3.2 Bandpass Filters Using Coaxial SIR

3.2.1 Synthesis Method of SIR-BPF Using Capacitive Coupling

Filters are among the most typical applications of SIR, and the design methods for such filters vary in accordance with the SIR structure. In this section, we exemplify a direct-coupled multistage bandpass filter consisting of a $\lambda_g/4$-type SIR as a unit resonant element. Though various coupling methods such as capacitive, inductive, and electromagnetic coupling can be adopted to this filter, here we employ a capacitive coupling method which is most suitable for compact design. When performing filter synthesis using distributed-element resonators like SIR, the resonators are usually expressed as electrically equivalent lumped-element components approximated at the resonance frequency. This enables the designer to directly apply an established synthesis method based on lumped-element circuitry. However, this lumped-element approximation is valid only in the frequency band near resonance. Thus, in designing a bandpass filter, one must keep in mind that although an accurate estimation can be obtained for frequency responses near pass-band, this approximation cannot be applied to calculate response at the stop-bands far from center frequency. This issue will be discussed further in Chap. 4 where we consider a BPF with wide stop-band characteristics.

A lumped-element expression of SIR can be obtained by calculating the susceptance slope parameters discussed in Sect. 2.4. Figure 3.4 illustrates the basic circuit configuration of an n-stage BPF composed of capacitively coupled shunt resonators. The synthesis method of this filter, originally proposed by S.B.Cohn [6], is slightly modified to meet with SIR structure. For simplicity, all inductors composing the filter are assumed to have identical inductance values, while the center angular frequency ω_0, number of stages n, element value g_j, relative bandwidth w, and input/output conductance G_A

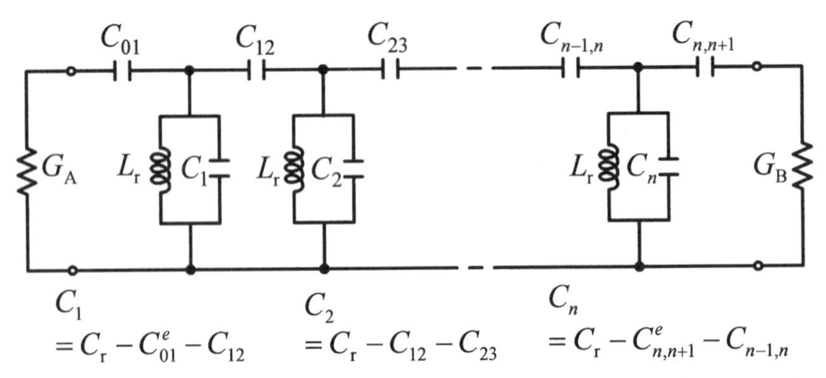

Fig. 3.4. Basic circuit configuration of capacitively coupled n-stage lumped-element resonator BPF

and G_B are to be assumed given as filter design parameters [7]. Under these conditions, coupling capacitance $C_{j,j+1}$ and shunt capacitance C_j can be expressed by the following equations.

Coupling capacitance:

$$C_{01} = \frac{J_{01}}{\omega_0 \sqrt{1-(J_{01}/G_A)^2}}, \tag{3.30a}$$

$$C_{j,j+1} = \frac{J_{j,j+1}}{\omega_0} \quad (j = 1 \text{ to } n-1), \tag{3.30b}$$

$$C_{n,n+1} = \frac{J_{n,n+1}}{\omega_0 \sqrt{1-(J_{n,n+1}/G_B)^2}}, \tag{3.30c}$$

where $J_{j,j+1}$: admittance inverter parameter [7] of a coupled section

$$J_{01} = \sqrt{\frac{G_A w}{g_0 g_0 \omega_0 L_r}}, \tag{3.31a}$$

$$J_{j,j+1} = \frac{w}{\omega_0 L_r \sqrt{g_j g_{j+1}}} \quad (j = 1 \text{ to } n-1), \tag{3.31b}$$

$$J_{n,n+1} = \sqrt{\frac{G_B w}{g_n g_{n+1} \omega_0 L_r}}. \tag{3.31c}$$

Shunt capacitance of resonators:

$$C_1 = C_r - C_{01}^e - C_{12}, \tag{3.32a}$$

$$C_j = C_r - C_{j-1,j} - C_{j,j+1} \quad (j = 2 \text{ to } n-1), \tag{3.32b}$$

$$C_n = C_r - C_{n,n-1}^e - C_{n-1,n}, \tag{3.32c}$$

where

$$C_r = 1/L_r \omega_0^2, \tag{3.33a}$$

$$C_{01}^e = C_{01}/\lfloor 1 + (\omega_0 C_{01}/G_A)^2 \rfloor, \tag{3.33b}$$

$$C_{n,n+1}^e = C_{n,n+1}/\lfloor 1 + (\omega_0 C_{n,n+1}/G_B)^2 \rfloor. \tag{3.33c}$$

Based on these fundamental equations, SIR is brought into the discussion as a resonator element. As with L_r in the above discussion, all SIR composing the filter are assumed to possess identical slope parameters. We adopt this restriction only for design convenience, as it is not always necessary. Furthermore, the resonance frequency of each SIR is adjusted by shortening resonator length in order to cancel a downward shift due to coupling capacitance. This adjustment length is represented in the following discussions by $\Delta \ell_j (j = 1 \text{ to } n)$.

Assuming that the basic structural parameters of SIR such as Z_1, Z_2, ℓ_1, ℓ_2, and slope parameter b_{SA} are previously determined, we obtain the following equation from (2.24) in Chap. 2.

$$\omega_0 L_r = \frac{1}{b_{SA}}$$

For $\ell_1 = \ell_2$, (2.21) gives $b_{\mathrm{SA0}} = \theta_{01} Y_2$. Therefore, the admittance inverter parameters $J_{j,j+1}$ are obtained as follows:

$$J_{01} = \sqrt{\frac{G_{\mathrm{A}} w \theta_{01} Y_2}{g_0 g_1}}, \tag{3.34a}$$

$$J_{j,j+1} = \frac{w \theta_{01} Y_2}{\sqrt{g_j g_{j+1}}}, \tag{3.34b}$$

$$J_{n,n+1} = \sqrt{\frac{G_{\mathrm{B}} w \theta_{01} Y_2}{g_n g_{n+1}}}. \tag{3.34c}$$

Focusing on the adjustment of resonator line length $\Delta\ell_j$, we recall from Sect. 3.1.2 that an equivalent transmission line length can replace a small capacitance at the open-end of a transmission line. In (3.32a), (3.32b) and (3.32c) the adjustment capacitance ΔC_j are represented as,

$$\Delta C_1 = C_{01}^{e} + C_{12}, \tag{3.35a}$$

$$\Delta C_j = C_{j-1,j} + C_{j,j+1} \quad (j = 2 \text{ to } n-1), \tag{3.35b}$$

$$\Delta C_n = C_{n,n+1}^{e} + C_{n-1,n}. \tag{3.35c}$$

Since $\omega_0 \Delta C_j << Y_2$, ΔC_j can be replaced by an equivalent transmission line length $\Delta\ell_j$ as

$$\Delta\ell_j = \frac{\lambda_{\mathrm{g}0}}{2\pi} \cdot \frac{\omega_0}{Y_2} \cdot \Delta C_j = \frac{v_{\mathrm{g}}}{Y_2} \cdot \Delta C_j \tag{3.36}$$

where v_{g} : phase velocity.

3.2.2 Design Examples and Performances

In this section we present the actual design procedures based on the design method described in Sect. 3.2.1, followed by the measured performance of an experimental filter based on this design. SIRs composing the filter in this discussion also assume basically the same shape, possessing a $\theta_1 = \theta_2$ structure. Filter design and fabrication procedures are as follows:

(i) Setting fundamental filter parameters based on given specifications,
(ii) Determining fundamental physical dimensions of the SIR,
(iii) Calculating coupling capacitance values,
(iv) SIR line length correction,
(v) Processing and assembly.

Notable points for each step are described as follows.

(a) Design Specifications and Fundamental Filter Parameters

The following specifications, assuming a mobile communication application, are presented as an example.

Center frequency	:	$f_0 = 900\,\text{MHz}$
Pass-band bandwidth	:	$W > 10\,\text{MHz}$
Pass-band insertion loss	:	$L_0 < 2\,\text{dB}$
Pass-band VSWR	:	< 1.2
Attenuation	:	$L_S > 80\,\text{dB}$ *at* $f_0 \pm 50\,\text{MHz}$
Resonator length	:	$< 40\,\text{mm}$

Fundamental filter parameters are determined to satisfy the above specifications while minimizing stage number n. One such example is shown below.

Filter response type	:	Chebyshev type[7]
Number of stages	:	$n = 6$
Relative bandwidth	:	$w = 0.022$
Pass-band ripple	:	$R = 0.01\,\text{dB}$

Filter element values g_j [7] are given as follows:

$$g_0 = 1.0000 \qquad g_1 = 0.7813$$
$$g_2 = 1.3600 \qquad g_3 = 1.6897$$
$$g_4 = 1.5350 \qquad g_5 = 1.4970$$
$$g_6 = 0.7098 \qquad g_7 = 1.1007$$

Insertion loss L_0 [7] at midband is calculated as,

$$L_0 \approx \frac{4.434}{wQ_0} \cdot \sum_{j=1}^{n} g_j\,(\text{dB}) = 1490/Q_0\,(\text{dB}). \tag{3.37}$$

Specifications restricting insertion loss to $L_0 < 2\,\text{dB}$, imply that the unloaded-Q is required to be.

$$Q_0 > 740.$$

(b) Determining Fundamental Physical Dimensions of the SIR

Considering the given resonator length and required unloaded-Q, the fundamental structure of the SIR, namely a_1, a_2, b, ℓ_1, and ℓ_2 in Fig. 3.1a, is determined. In this case also, we assume that $\theta_1 = \theta_2$. The resonator length is temporarily set as 35 mm in order to satisfy given specifications of under 40 mm. Assuming an air-cavity type SIR, normalized resonator length is obtained as follows:

$$L_{n0} = \ell_{\text{T}}/\ell_0 = 40/83.3 = 0.42.$$

R_Z is obtained from (2.14) or Fig. 2.4 both in Chap. 2 as

$$R_Z = \tan^2\theta_{01} = 0.11.$$

An approximate value of unloaded-Q can be obtained from Fig. 3.3 as

$$Q_0 = Q_{\text{C}} \approx 0.78 Q_{\text{CM}}.$$

Assuming the conductor material is copper, Q_C is given by (3.28) as

$$Q_{CM} \approx 2670b\sqrt{f_0} = 1975b \quad (b : \text{in cm}).$$

The Q value calculated from the above equation is an estimated value that is derived from insertion loss at the center frequency of the pass-band. In practical design, however, factors such as an increase in insertion loss at the edge of the pass-band, and degradation of unloaded-Q due to conductor surface conditions must be taken into account. Thus, considering an adequate margin of approximately 20%, the required unloaded-Q value is raised to $Q_0 > 1000$. This determines the outer conductor radius b as,

$$b = 0.75\,\text{cm} = 7.5\,\text{mm}.$$

Figure 3.3 suggests that Z_1 should be between $75\,\Omega$ to $90\,\Omega$ to obtain a maximum Q value for $R_Z = 0.11$. Consequently, the inner conductor radius is determined as,

$$a_1 = 2.0\,\text{mm}, \qquad a_2 = 6.5\,\text{mm}.$$

Thus from the above discussion, the SIR electrical parameters are obtained as follows:

$$\begin{aligned} Z_1 &= 60\ln(b/a_1) = 79.3\,\Omega, \\ Z_2 &= 60\ln(b/a_2) = 8.6\,\Omega, \\ R_Z &= Z_2/Z_1 = 0.108. \end{aligned}$$

Having determined the physical dimensions of SIR, the discontinuity capacitance at the step junction of the center conductor and the open-end fringing capacitance can be calculated in accordance to the discussions Sect. 3.1.2, giving

$$\begin{aligned} &\text{Discontinuity capacitance} \quad &: C_d \approx 0.4\,\text{pF} \\ &\text{Open-end fringing capacitance} &: C_f \approx 0.5\,\text{pF}. \end{aligned}$$

A fine-tuning capacitance mechanism, adjustable between 0.1 pF to 0.5 pF, is loaded at the open-end to absorb the processing variance of resonator physical dimensions. Correction of line length $\Delta\ell$ due to the step junction and open-end capacitance values can be estimated from (3.7), and consequently the line length of the center conductor is given as,

$$\begin{aligned} \ell_1 &= 34.0/2 = 17.0\,\text{mm}, \\ \ell_2 &= \ell_1 - \Delta\ell = 13.2\,\text{mm}. \end{aligned}$$

Thus the fundamental SIR structure, as shown in Fig. 3.5, is determined from the above discussion. $\Delta\ell_j$ in the figure indicates the line length correction due to the coupling capacitors.

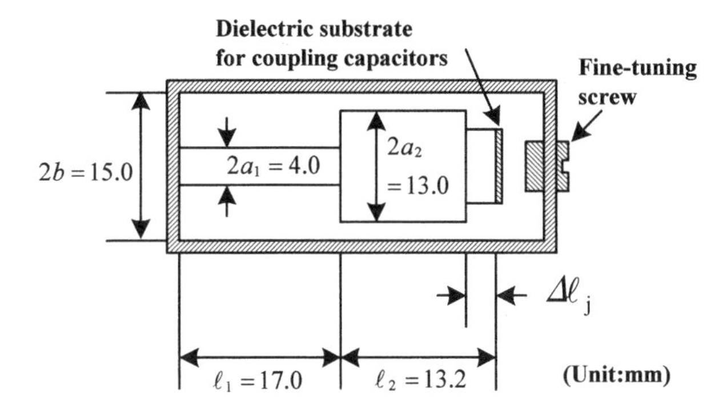

Fig. 3.5. Physical dimensions of the unit SIR for the experimental BPF

(c) Calculating coupling capacitance

Once the slope parameters and inverter parameters are known, coupling capacitance can be calculated from (3.30a), (3.30b) and (3.30c). Assuming $\theta_1 = \theta_2$,

$$b = \theta_1 Y_2 = 0.037\ \mathrm{S}.$$

Let input/output impedance be $50\,\Omega$, then $G_{\mathrm{A}} = G_{\mathrm{B}} = 1/50 = 0.02\,\mathrm{S}$, and $w = 0.0222$. Accordingly, the inverter parameters are obtained from (3.34a), (3.34b) and (3.34c) as follows:

$$\begin{aligned} J_{01} = J_{67} &= 4.56 \times 10^{-3}\ \mathrm{S}, \\ J_{12} = J_{56} &= 0.787 \times 10^{-3}\ \mathrm{S}, \\ J_{23} = J_{45} &= 0.535 \times 10^{-3}\ \mathrm{S}, \\ J_{34} &= 0.505 \times 10^{-3}\ \mathrm{S}. \end{aligned}$$

From the above results, coupling capacitance values are derived as,

$$\begin{aligned} C_{01} = C_{67} &= 0.831\ \mathrm{pF}, \\ C_{12} = C_{56} &= 0.139\ \mathrm{pF}, \\ C_{23} = C_{45} &= 0.095\ \mathrm{pF}, \\ C_{34} &= 0.089\ \mathrm{pF}. \end{aligned}$$

In this example, an interdigital type and a gap-type capacitor are adopted for the coupling capacitance in order to realize highly accurate capacitance values, as shown in Fig. 3.6. Interdigital capacitors are applied to input and output coupling requiring large capacitance values, while gap capacitors are applied to interstage couplings accompanying relatively small capacitance values. Capacitance values of such capacitors are dependent on physical structure and the applied dielectric material, as in Fig. 3.7 where we exemplify both types based on a substrate with a relative dielectric constant ε_{r} of 2.7 and thickness H of 0.3 mm. Using these design charts, the finger length ℓ of the interdigital capacitor and gap spacing g of the gap capacitor are obtained as follows:

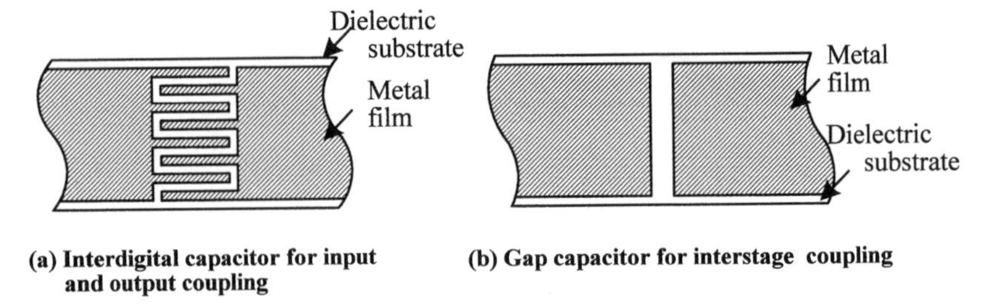

Fig. 3.6. Examples of coupling capacitor. (**a**) Interdigital capacitor suitable for I/O coupling. (**b**) Gap capacitor suitable for interstage coupling

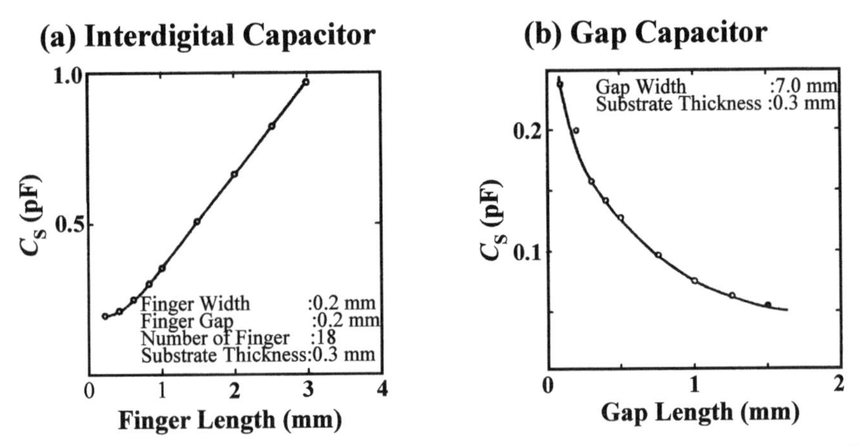

Fig. 3.7. Design examples of coupling capacitors. (**a**) Interdigital capacitor. (**b**) Gap capacitor

Input and output coupling : $\ell = 2.68\,\text{mm}$ (C_{01}, C_{67}).
Interstage coupling : $g = 0.50\,\text{mm}$ (C_{12}, C_{56}),
$g = 0.80\,\text{mm}$ (C_{23}, C_{45}),
$g = 0.90\,\text{mm}$ (C_{34}).

Both types of capacitors can be integrated on a single substrate by precise photoetching technology.

(d) Resonator Length Correction

With the coupling capacitance values determined, adjustment of the capacitance value ΔC_j can be obtained from (3.35a), (3.35b) and (3.35c) as follows:

$$\Delta C_1 = \Delta C_6 = 0.924\text{ pF},$$
$$\Delta C_2 = \Delta C_5 = 0.234\text{ pF},$$
$$\Delta C_3 = \Delta C_4 = 0.184\text{ pF}.$$

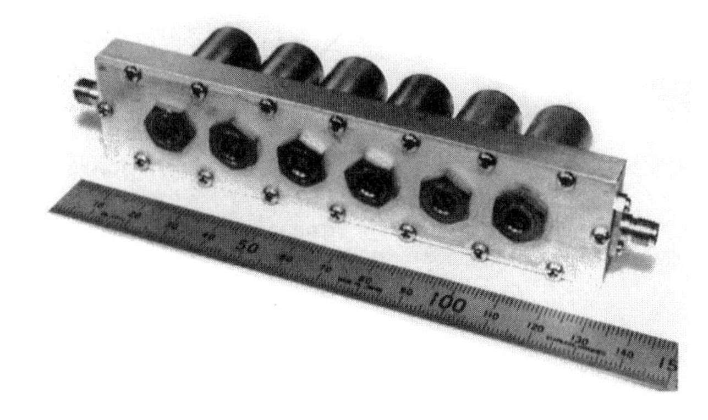

Fig. 3.8. Photograph of the experimental BPF

Substituting these results into (3.36), we obtain

$$\begin{aligned}
\Delta\ell_1 &= \Delta\ell_6 = 2.39 \text{ mm},\\
\Delta\ell_2 &= \Delta\ell_5 = 0.61 \text{ mm},\\
\Delta\ell_3 &= \Delta\ell_4 = 0.48 \text{ mm}.
\end{aligned}$$

Having integrated a fine-tuning mechanism at the open- ends of each resonator, the above results are approximated to obtain $\Delta\ell_j$ as follows:

$$\Delta\ell_1 = \Delta\ell_6 = 2.4 \text{ mm},$$
$$\Delta\ell_2 = \Delta\ell_3 = \Delta\ell_4 = \Delta\ell_5 = 0.6 \text{ mm}.$$

(e) Fabrication and Electrical Performances

The above design parameters were employed in the fabrication and adjustment of an experimental filter. Copper was chosen for the conductor material, and flanged cans were employed for the outer conductor of the resonators. Two interdigital capacitors and five gap capacitors were integrated on a dielectric substrate possessing the above-mentioned physical characteristics, and this substrate was attached to the open-end of the resonators with screws. In addition, mechanical tuning screws for fine adjustment were installed on the side-walls of the outer conductor opposed to the open-end of the resonators. Figures 3.8 and 3.9 show photographs of the filter appearance and integrated coupling capacitors. Post-fabrication adjustments generally require both coupling adjustment and fine-tuning of resonance frequency. In this design example, the coupling capacitors are realized by a precise photoetching process, thus making it possible to neglect such coupling adjustments. For the remaining resonance frequency adjustment, fine-tuning of the resonators was performed by a conventional step-by-step method [7]. Figure 3.10 shows attenuation and return loss characteristics of the experimental filter, where solid and dotted lines represent the measured and designed data, respectively. The figure suggests that the measured results agree well with

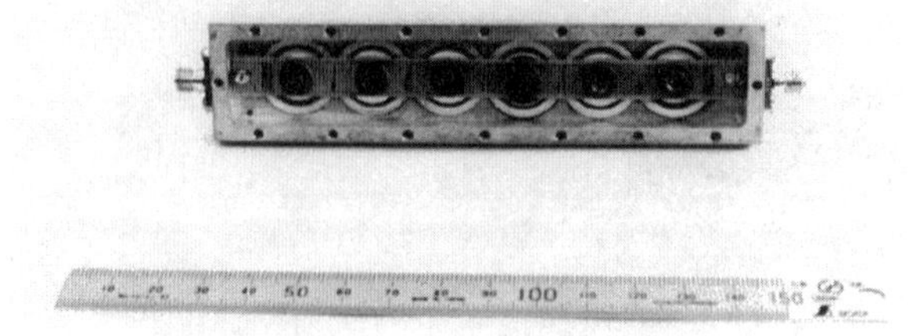

Fig. 3.9. Photograph of the coupling circuit structure (Cover plate removed)

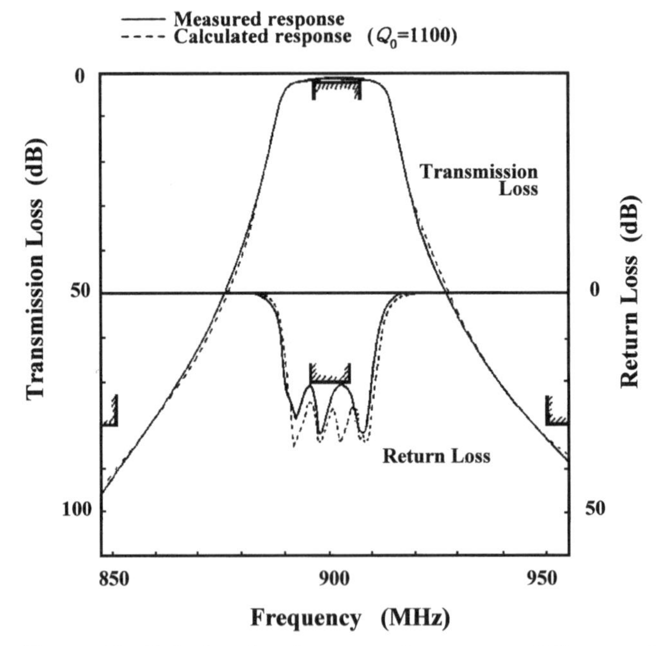

Fig. 3.10. Calculated and measured responses of the experimental BPF

design, especially for transmission characteristics, thus verifying the validity of the design method described above. Insertion loss at midband and band-edge was designed as 1.5 dB and 2.0 dB, respectively. Actual measurements showed 1.45 dB at midband and 1.65 dB at band-edge, these results satisfying target specifications. Furthermore, the unloaded-Q estimated from insertion loss measured at center frequency was calculated to be 1100, while direct measurement of a single resonator showed an unloaded-Q of 1200. This degradation is assumed to be loss due to the dielectric substrate mounted on the open-end of the resonators.

3.3 Double Coaxial SIR (DC-SIR)

3.3.1 Advantages of DC-SIR

In Sects. 3.1 and 3.2, we demonstrated the basic structure of a $\lambda_g/4$-type SIR and discussed its attractive properties of small size and high Q value. From a practical point of view, however, there still remains room for improvement of its basic structure. For example, the illustration in Fig. 3.5 suggests that antivibration characteristics of this SIR structure are insufficient for practical use due to a heavy weight at the end of the center conductor. In addition, we notice that the inner area of the center conductor contains no electromagnetic field, meaning that the space is not effectively used for increasing the Q value. As is apparent from the definition of unloaded-Q, the effective use of space within the resonator becomes critical for the enhancement of the Q value, because electromagnetic energy within an air-cavity type resonator is stored in the space enclosed by the inner and outer conductors. To overcome these problems, a DC-SIR (double coaxial SIR) which adopts a double coaxial structure at the tip of the center conductor as shown in Fig. 3.11 has been devised [8]. This structure significantly improves antivibration characteristics while providing increased design flexibility as compared to a basic SIR. Most important, this DC-SIR is expected to enable further miniaturization and a higher Q value, thus proving to be a highly practical resonator structure.

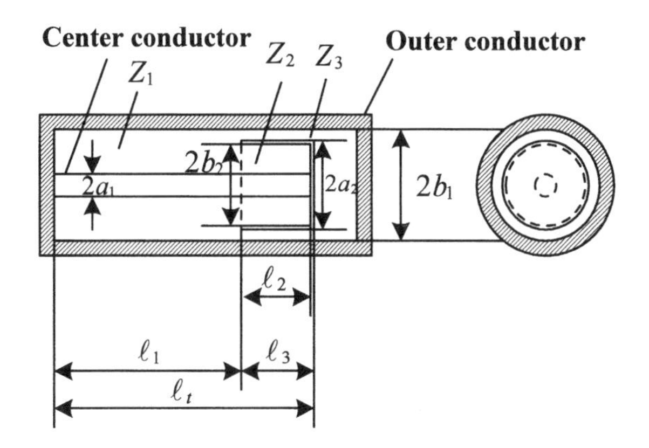

Fig. 3.11. Basic structure of a double coaxial SIR (DC-SIR)

3.3.2 Resonance Condition and Unloaded-Q

The DC-SIR has a short-circuited structure at one end of the center conductor, and an open-circuited structure at the other. From an electrical point of view, the DC-SIR is composed of three transmission-lines I, II, and III, each with different characteristic impedance Z_1, Z_2 and Z_3. Figure 3.12 shows an equivalent circuit of a DC-SIR expressed by transmission lines [9]. Let the

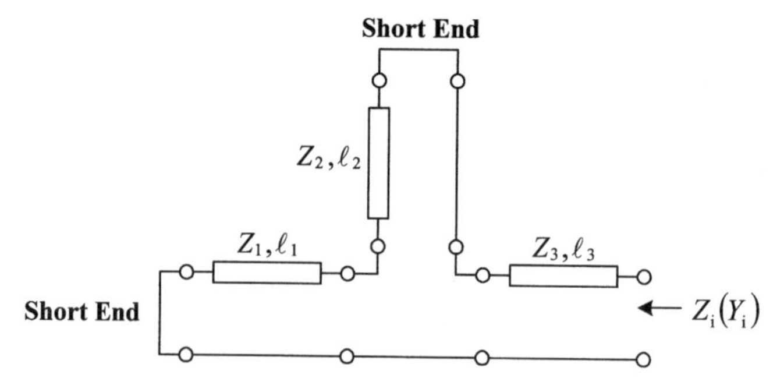

Fig. 3.12. An equivalent circuit of a DC-SIR using transmission lines

input impedance and admittance observed from the open-end of transmission line III be Z_i and Y_i, then we obtain,

$$\begin{aligned} Z_\mathrm{i} &= 1/Y_\mathrm{i} \\ &= -\mathrm{j}Z_3 \cdot \frac{\tan\beta\ell_1/R_{Z1} + \tan\beta\ell_2/R_{Z2} + \tan\beta\ell_3}{(\tan\beta\ell_1/R_{Z1} + \tan\beta\ell_2/R_{Z2})\tan\beta\ell_3 - 1} \end{aligned} \tag{3.38}$$

where

$$R_{Z1} = Z_3/Z_1, \quad R_{Z2} = Z_3/Z_2 \quad \text{and} \quad \beta = 2\pi/\lambda_{\mathrm{g}0}.$$

Resonance conditions are derived from (3.38) as $Y_\mathrm{i} = 0$, thus

$$(R_{Z2}\tan\beta\ell_1 + R_{Z1}\tan\beta\ell_2)\tan\beta\ell_3 = R_{Z1} \cdot R_{Z2}. \tag{3.39}$$

Slope parameters of DC-SIR can be obtained from the input admittance Y_i using (2.19). Let the electrical line length of each transmission line at resonance be θ_{01}, θ_{02} and θ_{03}, we obtain

$$\begin{aligned} b_\mathrm{s} &= \frac{\omega_0}{2} \cdot \left.\frac{\mathrm{d}B}{\mathrm{d}\omega}\right|_{w=w_0} \\ &= \frac{Y_3}{2} \cdot \left[\theta_{03} + \tan^2\theta_{03} \cdot \left(\frac{\theta_{01}}{R_{Z1}\cos^2\theta_{01}} + \frac{\theta_{01}}{R_{Z2}\cos^2\theta_{02}}\right)\right]. \end{aligned} \tag{3.40}$$

Thus, an equivalent circuit expression can be derived from the above result.

Furthermore, the unloaded-Q of a DC-SIR can be obtained by applying the method discussed in Sect. 3.1.3 of this chapter as follows:

$$\begin{aligned} Q_0 &= Q_\mathrm{C} \\ &= \frac{2b_1}{\delta} \cdot \frac{C_1}{D_1}, \end{aligned} \tag{3.41}$$

where

$$\begin{aligned} C_1 &= A_1 \ln\left(\tfrac{b_1}{a_1}\right) + A_2B_1 \ln\left(\tfrac{b_2}{a_1}\right) + A_3B_2 \ln\left(\tfrac{b_1}{a_2}\right), \\ D_1 &= A_1\left(1 + \tfrac{b_1}{a_1}\right) + A_2B_1\left(\tfrac{b_1}{a_1}\right)\left(1 + \tfrac{b_2}{a_1}\right) + A_3B_2\left(1 + \tfrac{b_1}{a_2}\right) \\ &\quad + \left(\tfrac{8\pi b_1}{\lambda_0}\right)\left[\ln\left(\tfrac{b_1}{a_1}\right) + B_1 \ln\left(\tfrac{b_2}{a_1}\right)\right], \end{aligned}$$

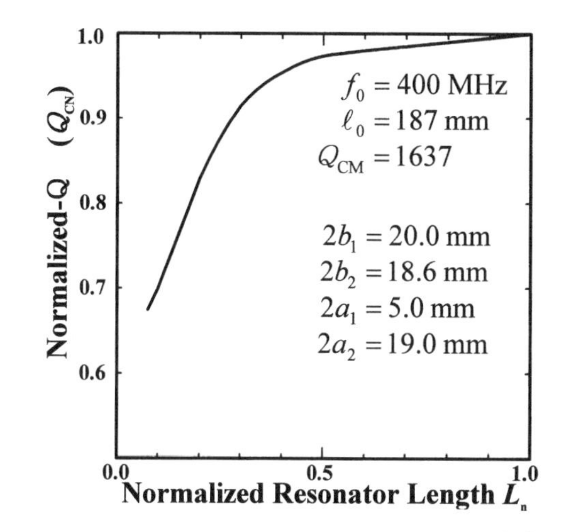

Fig. 3.13. Calculated results of unloaded-Q of DC-SIR

$$\begin{aligned} A_1 &= 2\theta_{01} + \sin\theta_{01}, \quad A_2 = 2\theta_{02} + \sin 2\theta_{02}, \\ A_3 &= 2\theta_{03} - \sin\theta_{03}, \\ B_1 &= \cos^2\theta_{01}/\cos^2\theta_{02}, \quad B_2 = \cos^2\theta_{01}/\sin^2\theta_{03}. \end{aligned}$$

Using these analytical results, the relationship between resonator length and unloaded-Q is examined. Figure 3.13 shows the unloaded-Q of a DC-SIR possessing physical dimensions of $2b_1 = 20.0\,\text{mm}$, $2b_2 = 18.6\,\text{mm}$, $2a_1 = 5.0\,\text{mm}$, and $2a_2 = 19.0\,\text{mm}$, applying copper as the conductor material. The horizontal and vertical axes, respectively, indicate resonator length and unloaded-Q, both normalized by the length ($\ell_0 = 187\,\text{mm}$) and Q value ($Q_{\text{CM}} = 1637$) of a $\lambda_g/4$ UIR. Calculations are based on a fixed resonance frequency of 400 MHz, thus the line length ratio ℓ_2/ℓ_1 increases in accordance with a decrease in total line length.

At a fixed resonance frequency, the unloaded-Q of a conventional UIR with loaded capacitors is approximately proportional to line length [10]. However, in the case of a DC-SIR, the above results indicate that degradation of the Q value due to a decrease in resonator length is very small.

The example in the figure claims Q values of 96% and 84% even when the length of the DC-SIR is shortened to 50% and 20%, respectively. One advantage of the SIR is this feature of maintaining a high Q value even when resonator length is shortened.

3.3.3 400 MHz-Band High-Power Antenna Duplexer

The antenna duplexer is an important device that enables simultaneous (duplex) operation of the receiver and transmitter in wireless communication. In a duplex wireless communication employed frequency division multiplex

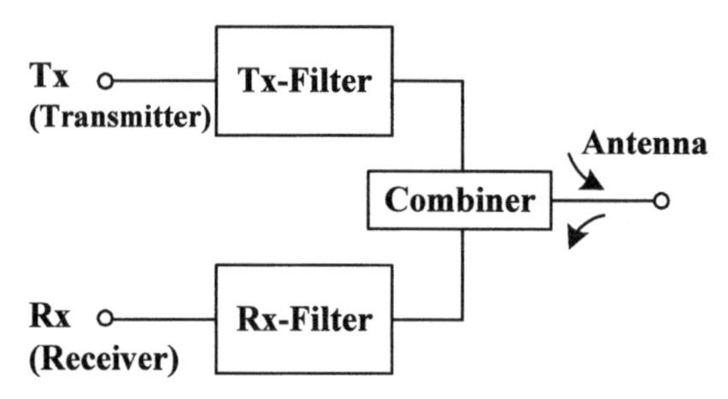

Fig. 3.14. Basic configuration of a duplexer

(FDM) system, the frequency bands of the Tx and Rx signals are assigned apart from one other, thus the duplexer is required to separate the Tx and Rx waves and share a single antenna for transmitter and receiver. A conventional duplexer, as shown in Fig. 3.14, is composed of a Tx-filter, an Rx-filter, and a combining circuit. A combining circuit is a phase adjustment circuit that is inserted to reduce mutual influences of the Tx and Rx-filters, usually consisting of reactance elements such as transmission lines, inductors, and capacitors.

Basic Tx-Rx filter combinations are BPF-BPF, BPF-BEF and BEF-BEF. When designing a wireless communication system, the most suitable configuration is chosen by taking into account the transmitter signal bandwidth and separation of the Tx and Rx frequency bands. For example, a BPF-BPF combination is often applied for a 900 MHz-band mobile communication system because the transmission bandwidth and frequency separation of the Tx and Rx-bands are approximately 15 MHz and 150 MHz, respectively. On the other hand, mobile communication systems utilizing a 400 MHz-band usually adopt 2 or 3 MHz for transmission bandwidth and 10 MHz for Tx and Rx-band separation, thus applying a BEF-BEF configuration for the reduction of insertion losses at the pass-band.

Issues taken into consideration when designing an antenna duplexer include minimizing influence between Tx- and Rx-filters, securing the interference characteristics of total radio system, and suppressing spurious responses.

As a design example, here we describe a 400 MHz-band high-power duplexer realized by utilizing DC-SIR [8]. The design specifications for this specific application, which we later describe in detail, require an unloaded-Q >1000 and ℓ_T <60 mm, and according to these restrictions a DC-SIR structure was chosen as $2b_1 = 20.0$ mm, $2b_2 = 18.6$ mm, $2a = 5.0$ mm, $2a_2 = 19.0$ mm, and $\ell_T = 50.0$ mm. Experimental data on resonance frequency and unloaded-Q value as a function of line length ℓ_3 are shown in Figure 3.15 and 3.16, respectively. Figure 3.15 suggests that the resonance frequency for a resonator of fixed total length ℓ_T can be controlled between 350 MHz to 500 MHz by changing line length ℓ_3. Theoretical estimations of resonance frequency given by (3.39) do not exactly match experiment results

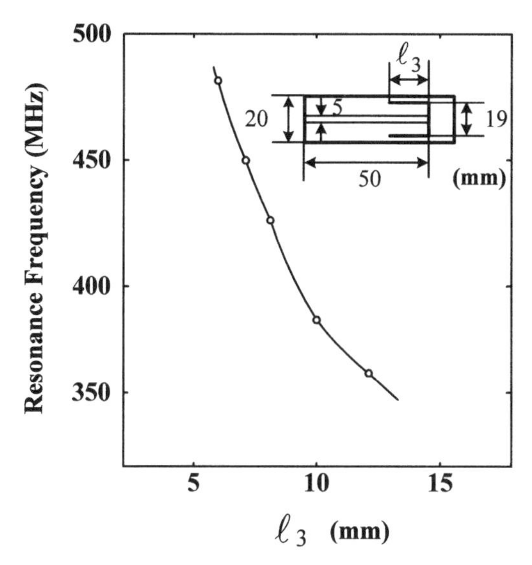

Fig. 3.15. Experimental results of resonance frequency of DC-SIR

because the equation neglects effects due to discontinuities of the DC-SIR. Closer agreement was obtained by considering the fringing capacitance at these discontinuities. Measured unloaded-Q values were about 80% of the theoretical values calculated from (3.41) due to effects of fabrication techniques such as surface finishing and electroplating.

As shown in Fig. 3.14, two separate filters are required to compose an antenna duplexer, and thus the design specifications are also given separately for both Tx- and Rx-filters. Specifications of each filter considered in this section are shown below.

Tx-filter

Center frequency	: f_0	$= 410.7\,\text{MHz}$
Pass-band bandwidth	: W	$= 3.3\,\text{MHz}$
Pass-band insertion loss	: L_0	$< 1.7\,\text{dB}$
Pass-band VSWR	:	< 1.3
Attenuation	: L_S	$> 60\,\text{dB}$ *at* 416 MHz
		$> 18\,\text{dB}$ *at* 404 MHz
Handling power	:	$> 30\,\text{W}$

Rx-filter

Center frequency	: f_0	$= 417.7\,\text{MHz}$
Pass-band bandwidth	: W	$= 3.3\,\text{MHz}$
Pass-band insertion loss	: L_0	$< 3.0\,\text{dB}$
Pass-band VSWR	:	< 1.5
Attenuation	: L_S	$> 55\,\text{dB}$ *at* 412.3 MHz

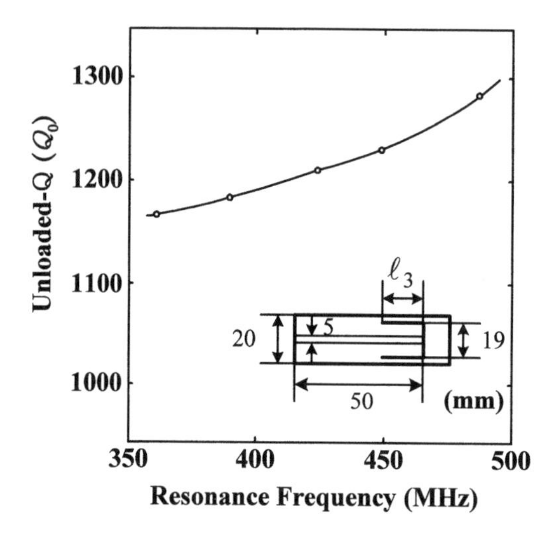

Fig. 3.16. Measured data of unloaded-Q of the experimental DC-SIR

The synthesis method described in Sect. 3.2 was applied to generate a filter design that could satisfy the specifications listed above. From these results we concluded that a 5-stage bandpass filter could be designed to realize the Rx-filter, while for the Tx-filter, a bandpass filter configuration was insufficient due to high losses. Thus, a combination of a 2-stage bandpass and a 4-stage band-elimination filter was adopted to realize the Tx-filter. Having no generalized design criteria for a band-elimination filter with a wide stop-band, a computer optimization technique for RF circuits was introduced to simultaneously satisfy stop-band attenuation and pass-band insertion losses in the filter design.

Figure 3.17 shows a photograph of the experimental duplexer. Physical dimensions of the DC-SIR unit measured 20 mm(diameter) × 55 mm(length), and an unloaded-Q of approximately 1200 was obtained. Measured transmission characteristics are illustrated in Fig. 3.18. Pass-band insertion losses were below 1.5 dB and 2.6 dB (including a 0.5 dB input cable loss) for Tx-

Fig. 3.17. Photograph of the 400 MHz-band duplexer using DC-SIRs

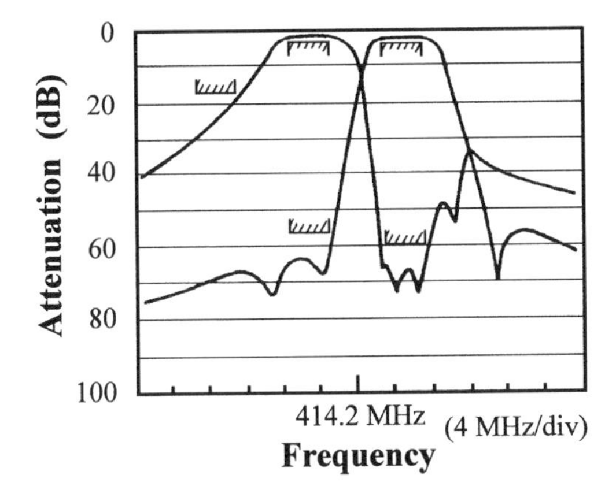

Fig. 3.18. Measured response of the experimental duplexer

and Rx-filters. The Tx-filter has no spurious response up to 3 GHz, proving a wide stop-band compared with conventional duplexers using $\lambda_g/4$ UIR or helical resonators. The Handling power of the Tx-filter proved to be more than 50 W, and a good temperature stability (less than 8 ppm/deg) was achieved by applying a compensation technique utilizing two metal materials with different thermal expansion coefficients such as copper and iron.

3.4 Dielectric Coaxial SIR

3.4.1 Dielectric Materials and Features of Dielectric Resonators

Dielectric materials are indispensable for circuit substrates in the microwave region. However, these substrate materials, such as alumina ceramics, have long been neglected as resonator materials due to their poor temperature stability. Ever since the development of novel ceramic materials possessing low-loss characteristics and adequate temperature stability in the early 1970s, much effort has been put into improving the microwave properties of ceramic materials. The progress of these materials has in turn accelerated their application to filters in the RF and microwave region.

Figure 3.19 illustrates typical structural variations of dielectric resonators. While typical dielectric resonators possess a cubic or cylinder-shaped structure and a TE or TM resonant mode, coaxial type resonators with a TEM resonant mode are also regarded as dielectric resonators in the following discussion. $TE_{01\delta}$ mode operated dielectric resonators are frequently applied to microwave and millimeter-wave filters for satellite communication and satellite broadcasting systems which require high performance filters with low insertion loss and excellent temperature stability. TEM-mode operated dielectric coaxial resonators are, in contrast, applied to filters and duplexers for mobile communication equipment due to a strong demand for miniaturization.

	Resonance Mode	Structure	Dimensions	Applicable Freq. Range
Diectric Resonator	$TE_{01\delta}$	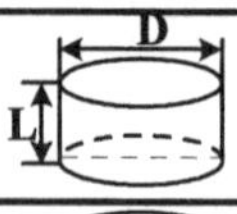 D, L	$D \cong \frac{C}{f_0\sqrt{\varepsilon_r}}$ $L/D \cong 0.4$	1 -- 100 GHz
Coaxial Resonator	TEM	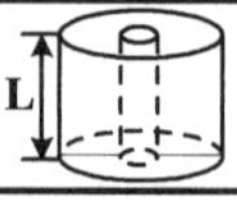 L	$L \cong \frac{C}{2f_0\sqrt{\varepsilon_r}}$ $(\lambda_g/2\ \text{Resonator})$	0.1 -- 5 GHz
Stripline Resonator	TEM	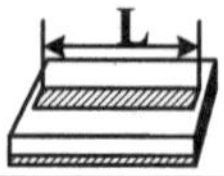 L	$L \cong \frac{C}{2f_0\sqrt{\varepsilon_r}}$ $(\lambda_g/2\ \text{Resonator})$	0.1 -- 100 GHz

C:Speed of light in free space

Fig. 3.19. Typical structures of RF/microwave resonator using dielectric materials

Basic properties required for the materials of such dielectric resonators are as follows:

(i) High dielectric constant,
(ii) Small loss-tangent ($\tan\delta_d$)
(iii) Good temperature stability.

As for requirement (i), the wavelength of an electromagnetic wave propagating through dielectric media is inversely proportional to $\sqrt{\varepsilon_r}$, and thus materials with a high dielectric constant contribute to the miniaturization of resonator as shown in Fig. 3.19. Requirement (ii) is directly related to the unloaded-Q (Q_0) of the resonator, which is understood from the relationship $1/Q_0 = 1/Q_C + \tan\delta_d$. The influence of Q_C is minimal for TE and TM mode dielectric resonators that possess no conductive electrodes. For requirement (iii), there is a necessity to reduce the temperature coefficient of the resonance frequency (τ_f), which can be achieved by lowering the temperature coefficient of dielectric constant (τ_s). An approximate relationship between τ_s and τ_f is given as follows:

$$\tau_f = -\frac{1}{2}\tau_s + \alpha_\ell \tag{3.42}$$

where α_ℓ : temperature expansion coefficient of the dielectric material.

A practical value of $\tau_f \approx 0$ ppm/°C requires a τ_s of approximately 20 ppm/°C for conventional ceramic materials with an a_ℓ of approximately 10 ppm/°C. Microwave properties of typical dielectric materials in practical use are summarized in Table 3.1 [11,12], where we see various ceramic materials with excellent temperature stability and high Q_d values. The discussion in this

Table 3.1. Material properties of typical microwave ceramics

Material	ε_r	Q_d	τ_f (ppm/°C)	Measured Freq. (GHz)
$(Mg_{0.95}Ca_{0.05})TiO_3$	21	16000	0	3
$Ba(Zn_{1/3}Ta_{2/3})O_3$	30	14000	0	12
$(ZrSn)TiO_4$	36	6500	0 ±2	7
$Ba(ZrTi)O_3$	37.5	6500	0	7.5
$BaOSm_2O_3TiO_3$	70–90	4000	0 ±4	2

chapter focuses on the dielectric coaxial SIR and its application to filters. The advantages of such an application are summarized as follows:

(i) Reduction of resonator length proportional to $1/\sqrt{\varepsilon_r}$,
(ii) Shaping and sintering flexibility of ceramic materials for various structures,
(iii) Antivibration characteristics,
(iv) Excellent stability against temperature and humidity,
(v) Cost-effective structures and manufacturing processes.

3.4.2 Basic Structure and Characteristics of Dielectric Coaxial Resonator

Figure 3.20 illustrates the structural variations of a dielectric coaxial resonator. Figure 3.20a indicates the conventional resonator (UIR) possessing uniform transmission-line impedance, while Figs. 3.20b and c show resonators with a stepped-impedance structure. The resonator shown in (b) employs conductors of round profile, with a uniform and step junction diameter for inner and outer conductors, respectively [13]. A contrary structure, of which the outer conductor possesses a uniform diameter and the inner conductor a step junction, is also applicable. The example shown in (c) has a structure aimed to improve unloaded-Q while simultaneously reducing size [14]. Improvement of unloaded-Q can be achieved by applying a design that allows maximum resonator volume, thus a structure possessing a square profile is preferable as compared to a round one. This structure is also valid for a filter assembly, where a square profile is far more convenient for positioning and soldering. From the above reasons, dielectric SIRs with square-shaped outer conductors are often adopted for practical devices. As previously described, miniaturization of the SIR requires the reduction of the impedance ratio R_Z. Figure 3.21 compares calculated results of characteristic impedance in the case of a round and square inner conductor, where the outer conductor is a square. These results are obtained by numerical analysis using a successive over-relaxation method. From these results we understand that to reduce the

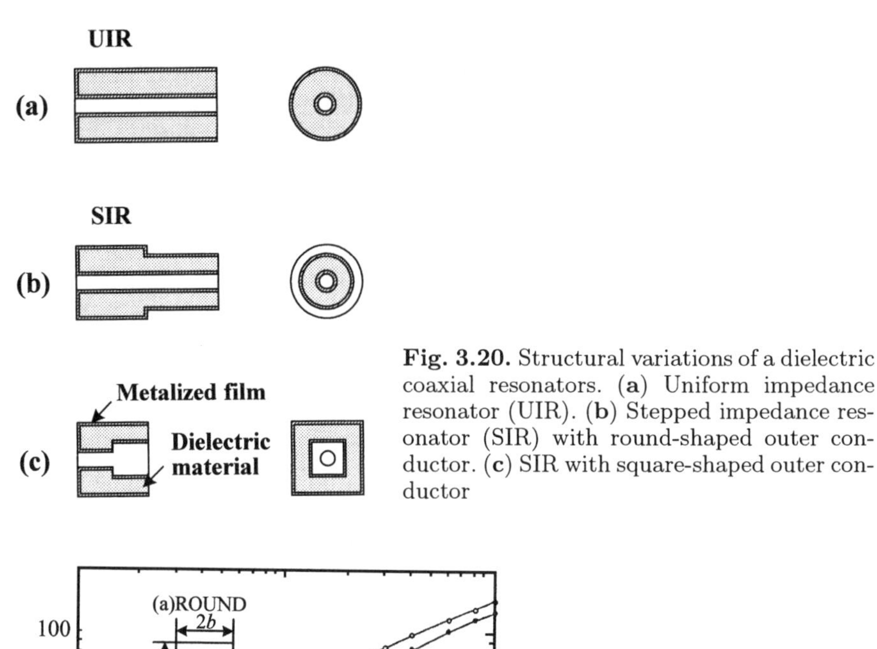

Fig. 3.20. Structural variations of a dielectric coaxial resonators. (**a**) Uniform impedance resonator (UIR). (**b**) Stepped impedance resonator (SIR) with round-shaped outer conductor. (**c**) SIR with square-shaped outer conductor

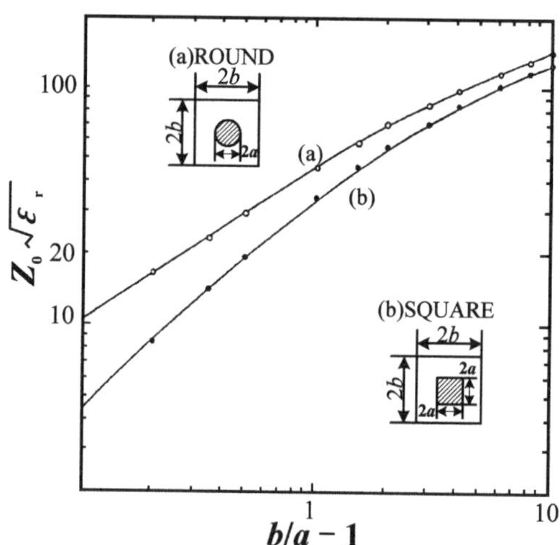

Fig. 3.21. Calculated results of characteristic impedance for a dielectric coaxial SIR

impedance ratio R_Z, the choice of a round conductor for Z_1 and a square conductor for Z_2 is desirable.

For an actual SIR structure, manufacturing restrictions yield a necessity to connect the two transmission lines of different impedance with a tapered section as illustrated in Fig. 3.22. Doing so results in a decrease in circuit loss because this tapered line mitigates the steep change in electromagnetic field distribution at the step junction. However, the introduction of this tapered section results in an increase in total resonator length. Figure 3.23 shows the resonator length as a function of the length of the tapered section. These results were obtained using a conventional microwave circuit simulator by approximating the tapered inner-conductor structure into N (N>10) cascaded

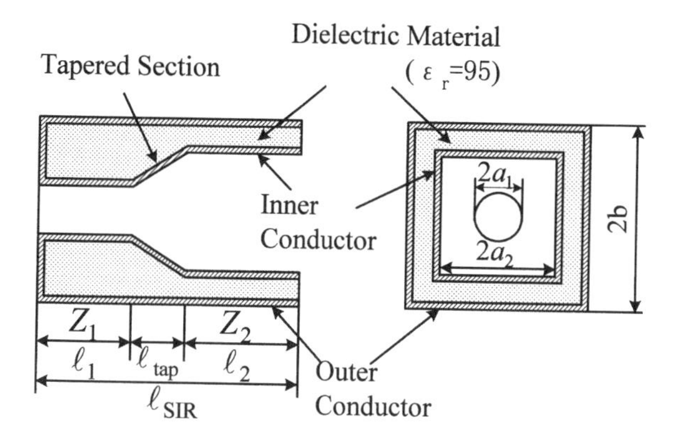

Fig. 3.22. Structure of coaxial SIR possessing internal tapered section

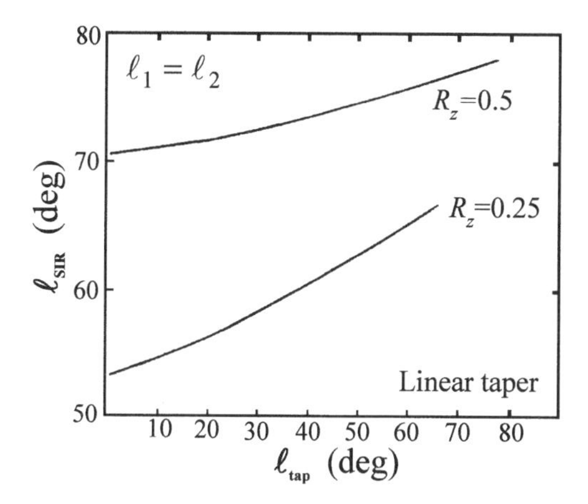

Fig. 3.23. Resonator length considering effects of tapered section

section of uniform diameter. It is evident from these results that the tapered section makes the resonator length longer. Hence, it is preferable to design the tapered section as short as manufacturing conditions will permit.

The unloaded-Q value can be obtained according to the method discussed in Sect. 3.1.3. However, deriving a closed form equation similar to the case of a round coaxial resonator becomes difficult due to a complicated resonator structure, and thus computer-assisted numerical analysis methods are often applied in the study of square-shaped SIR possessing tapered sections. Even so, calculated results often show poor agreement with measured values, and thus coaxial approximations based on square to round transformation, as shown in Fig. 3.24, usually suit practical use. Calculated Q_C examples of dielectric SIR obtained by this method are shown in Fig. 3.25.

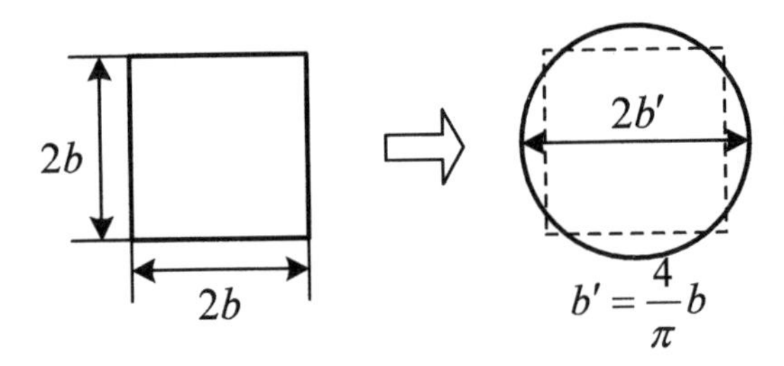

Fig. 3.24. The equivalent radius of square-shaped outer conductor

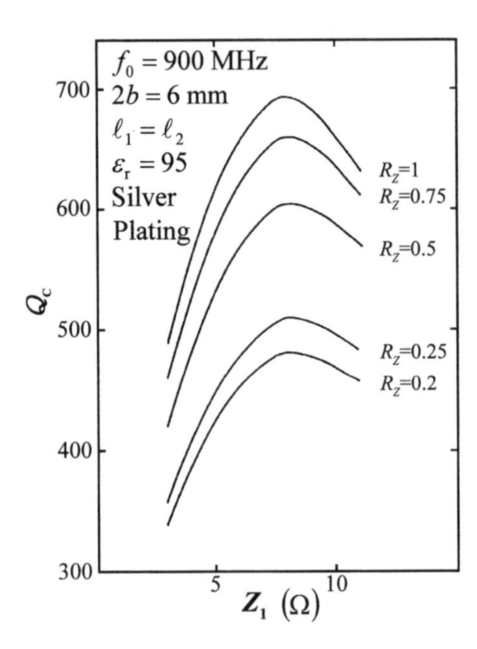

Fig. 3.25. Calculated results of Q_C-value of dielectric coaxial SIR

3.4.3 Design Example of Antenna Duplexer for Portable Radio Telephone

Applying the compact dielectric coaxial SIR structure discussed in the former section, an experimental antenna duplexer for an 800 MHz-band portable radio telephone terminal is designed and fabricated [14]. Specifications of the Tx- and Rx-filters are given as follows:

Tx-filter

Center frequency	: $f_0 = 927.5\,\mathrm{MHz}$
Pass-band bandwidth	: $W = 15\,\mathrm{MHz}$
Pass-band insertion loss	: $L_0 < 1.5\,\mathrm{dB}$
Pass-band VSWR	: < 1.5
Attenuation	: $L_S > 45\,\mathrm{dB}$ at Rx-Band

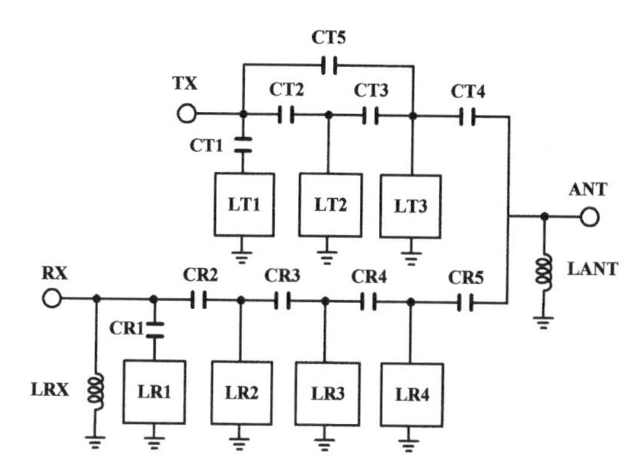

Fig. 3.26. Configuration of the experimental antenna duplexer

Rx-filter

Center frequency	: $f_0 \ = 827.5\,\mathrm{MHz}$
Pass-band bandwidth	: $W \ = 15\mathrm{MHz}$
Pass-band insertion loss	: $L_0 < 3.0\,\mathrm{dB}$
Pass-band VSWR	: < 1.5
Attenuation	: $L_\mathrm{S} > 45\mathrm{dB}$ at $\mathrm{Tx-Band}$

The frequency span between Tx- and Rx-bands for this application is wider than that of the previously discussed 400 MHz band duplexer example, and thus a BPF-based duplexer configuration is adopted, while adding an extra BEF stage to both Tx- and Rx-filters to secure attenuation characteristics. Figure 3.26 illustrates the basic configuration of the duplexer.

Filter simulation results concluded that an unloaded-Q value of at least 400 was required for each individual resonator, and thus the physical dimensions of the resonator, based on a dielectric material of $\varepsilon_\mathrm{r}=95$, were determined as follows.

Z_1 : Outer square conductor $2b = 6.0\,\mathrm{mm}$
Inner square conductor $2a_2 = 4.2\,\mathrm{mm}$
Z_2 : Outer square conductor $2b = 6.0\,\mathrm{mm}$
Inner round conductor $2a_1 = 1.6\,\mathrm{mm}$

Thus, from Fig. 3.21 we obtain

$$Z_1 = 9.62\,\Omega, \qquad Z_2 = 1.37\,\Omega,$$
$$R_Z = Z_2/Z_1 = 0.18.$$

The resonator length calculates to approximately 4.3 mm at 900 MHz, which is about half the size of a UIR applying the same dielectric material.

The loss tangent of the applied dielectric material is less than 2×10^{-4} at 1 GHz, and the measured unloaded-Q value of the resonator was approximately 420 at 900 MHz, which falls in close agreement with the results shown in Fig. 3.25. The actual appearance and a cross-sectional view of the manufactured SIR are shown in Fig. 3.27.

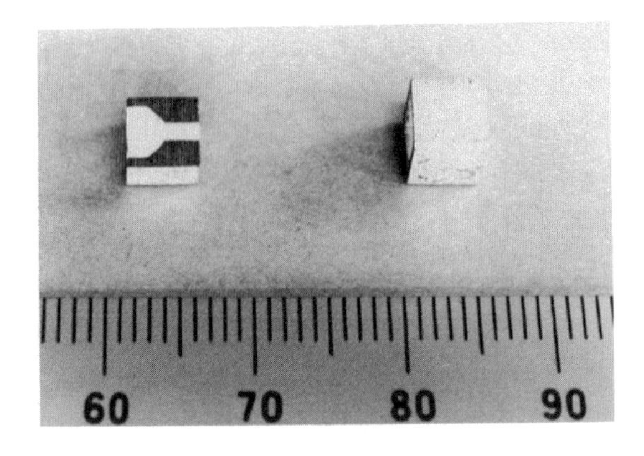

Fig. 3.27. Photograph of the manufactured dielectric coaxial SIR (right) and its cross-sectional cut-view (left)

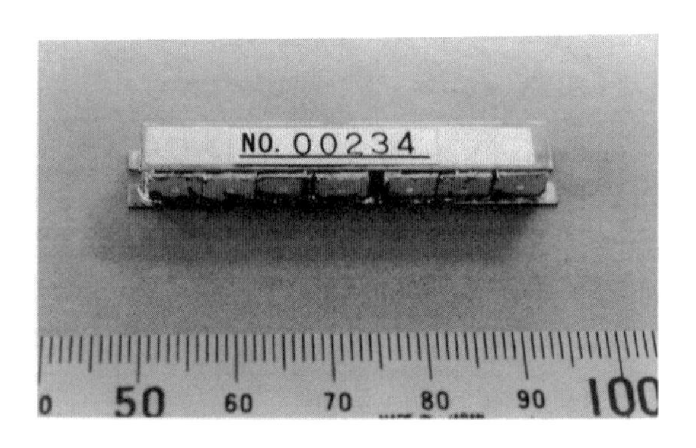

Fig. 3.28. Photograph of the experimental duplexer using dielectric coaxial SIRs

Figure 3.28 shows the experimental duplexer employing the above SIR. The Tx- and Rx-filters are respectively composed of three and four SIRs of basically same shape. Coupling circuits are integrated on a substrate of high dielectric constant to allow for a large coupling capacitance, this in correspondence to a more compact resonator design possessing a larger admittance slope parameter. Frequency tuning of each resonator is achieved by coarse adjustment of resonator length before filter assembly, while fine-tuning of the total device is performed by minute trimming of metalized film at the open-ends of the SIR. Physical dimensions of the experimental duplexer measure 45 mm(length)×6.5 mm(height)×8 mm(width), with a total volume of about 2.3 cm^{-3}. Weighing less than 6 g, this design reduces weight to half that of a conventional UIR-based duplexer.

Figure 3.29 shows the measured transmission responses of the experimental duplexer. Pass-band insertion losses were less than 2.5 dB for receiver and 1.3 dB for transmitter, while isolation levels of more than 60 dB in the transmitting band and 50 dB in the receiving band were achieved. These results meet the initial specifications of the antenna duplexer.

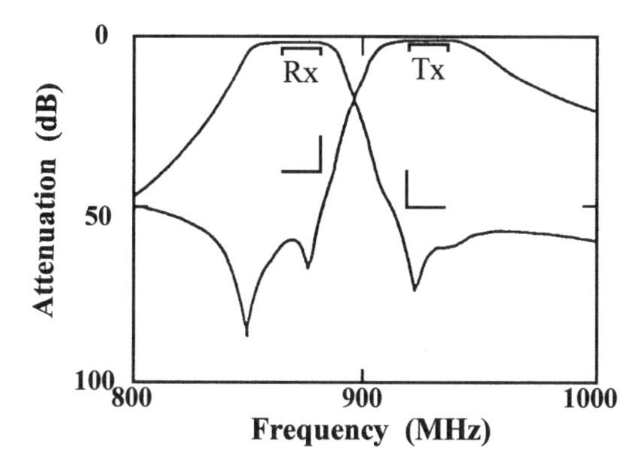

Fig. 3.29. Measured transmission responses of the experimental duplexer

3.4.4 Dielectric DC-SIR

From the discussions in Sect. 3.3, we understand that a double coaxial structure is desirable for further miniaturization of a conventional SIR [9]. As shown in Fig. 3.30c, the same approach can be applied to a dielectric coaxial resonator. Though the design process of a DC-SIR becomes somewhat difficult due to its complicated structure, once designed, ceramic molding and sintering processes can be applied as in the case of structures (a) and (b) in Fig. 3.30, thus making the DC-SIR suitable for mass-production. Figure 3.31 shows the basic structure and physical parameters of a dielectric DC-SIR.

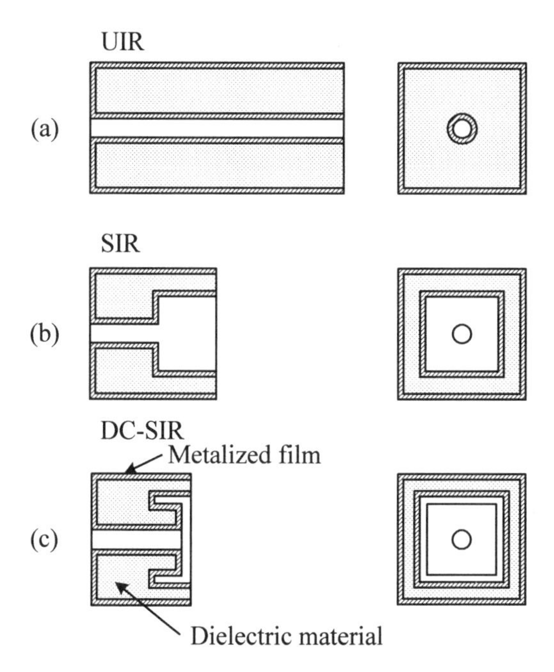

Fig. 3.30. Miniaturization of dielectric coaxial resonators. (**a**) Conventional UIR. (**b**) Square-shaped UIR. (**c**) Dielectric double-coaxial SIR (DC-SIR)

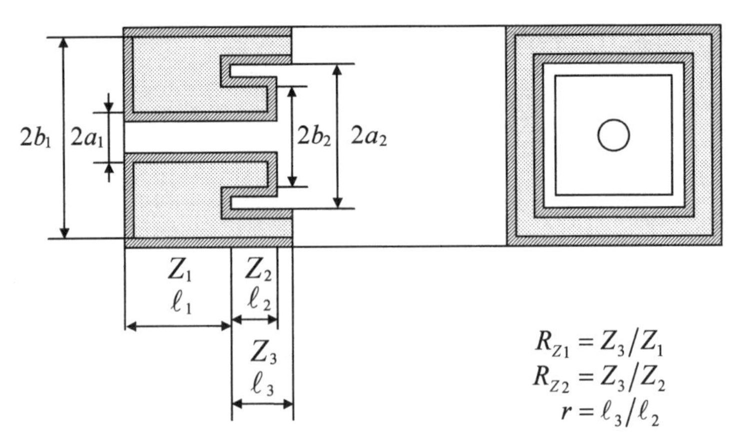

Fig. 3.31. Physical dimensions of a dielectric DC-SIR

The fundamental characteristics of this resonator structure can be analyzed by applying the techniques described in Sects. 3.3 and 3.4.2.

Neglecting the effects of the step discontinuity for simplicity, total resonator length $\ell_{\mathrm{DC-SIR}}$ is given as follows.

When $r = \ell_3/\ell_2 \leq 1$,

$$\begin{aligned}\ell_{\mathrm{DC-SIR}} &= \ell_1 + \ell_2 \\ &\cong \frac{1}{\beta}\tan^{-1}\left[R_{Z1}\left(1/\tan r\beta\ell_2 - \tan\beta\ell_2/R_{Z2}\right)\right] + \ell_2,\end{aligned} \tag{3.43}$$

and when $r = \ell_3/\ell_2 < 1$

$$\begin{aligned}\ell_{\mathrm{DC-SIR}} &= \ell_1 + \ell_3 \\ &\cong \frac{1}{\beta}\tan^{-1}\left[R_{Z1}\left(1/\tan r\beta\ell_2 - \tan\beta\ell_2/R_{Z2}\right)\right] + r\ell_2,\end{aligned} \tag{3.44}$$

where

$$R_{Z1} = Z_3/Z_2, \quad R_{Z2} = Z_3/Z_2$$
$$\beta = 2\pi\sqrt{\varepsilon_r}/\lambda_0, \quad \lambda_0 : \text{wavelength in free space.}$$

The normalized resonator length $L_{\mathrm{DC-SIR}}$ is defined as the resonator length ratio $\ell_{\mathrm{DC-SIR}}/\ell_{\mathrm{UIR}}$, where ℓ_{UIR} indicates the length of a dielectric coaxial UIR applying the same dielectric material. Figure 3.32 indicates $L_{\mathrm{DC-SIR}}$ as a function of L_1, i.e., l_1 in Fig. 3.31 normalized by ℓ_{UIR}.

A rough estimation of the unloaded-Q value can be obtained from (3.41) considering the square-to-round transformation of the coaxial structure described in Sect. 3.4.2. Figure 3.33 shows calculated results for $\varepsilon_{\mathrm{r}} = 95$ and $f_0 = 400\,\mathrm{MHz}$. From the figure we understand that the Q value of a DC-SIR has little dependence on Z_1 as compared to the data in Fig. 3.25. This

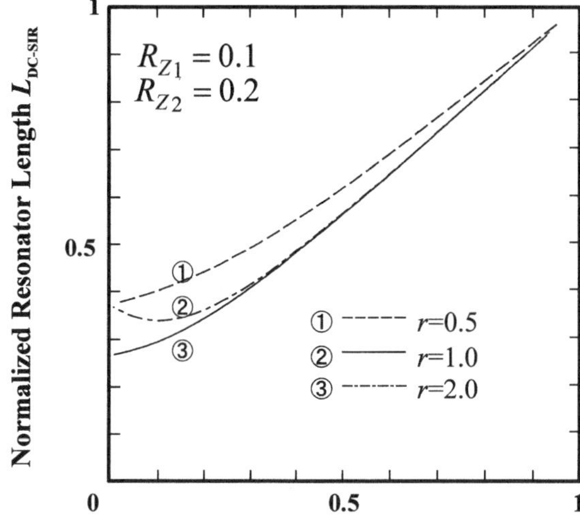

Fig. 3.32. Calculated results of resonator length

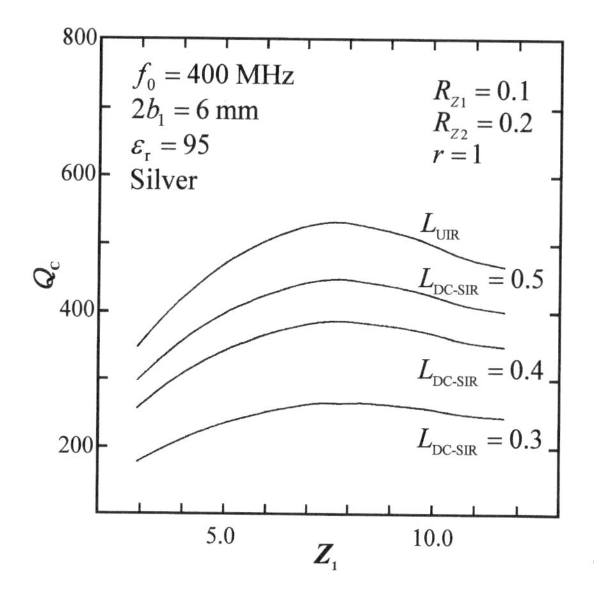

Fig. 3.33. Calculated results of unloaded-Q

understanding is based on an assumption that TEM mode is the dominant resonance mode of a DC-SIR. However, when $2b_1 > \ell_{DC-SIR}$, this dominant electromagnetic field is disturbed by higher TE or TM modes excited in the resonator, thus resulting in a remarkable degradation of unloaded-Q. Thus, to escape such effects, it is preferable to design a DC-SIR under the condition of $2b_1 > \ell_{DC-SIR}$. For example, in the case of $2b_1 = 6\,\text{mm}$, $\varepsilon_r = 95$ and $L_{DC-SIR} = 0.3$, the applicable frequency range of a DC-SIR is below

Fig. 3.34. Photograph of the manufactured dielectric DC-SIR (right) and its section (left)

400 MHz. From this discussion we understand that the DC-SIR structure is most effective in the lower frequency bands where the physical length of dielectric UIR and conventional SIR prove to be too long for precise and cost-effective manufacturing. This further implies that a DC-SIR structure provides a means to expand the applicable frequency region of dielectric coaxial resonators.

In the following section we explain an experimental DC-SIR designed at 400 MHz applying a dielectric material of $\varepsilon_r = 95$. Figure 3.34 shows a photograph of this DC-SIR manufactured by molding and sintering of ceramic materials. The physical dimensions are as follows,

$$\begin{aligned}
2a_1 &= 1.4\,\text{mm}, & 2a_2 &= 4.9\,\text{mm},\\
2b_1 &= 6.0\,\text{mm}, & 2b_2 &= 3.9\,\text{mm},\\
\ell_1 &= 3.1\,\text{mm}, & \ell_2 &= 2.4\,\text{mm},\\
\ell_3 &= 3.0\,\text{mm}, & \ell_T &= \ell_1 + \ell_3 = 6.1\,\text{mm}.
\end{aligned}$$

The electrical parameters of the resonator are given as,

$$Z_1 = 10.0\,\Omega, \qquad Z_2 = 7.5\,\Omega, \qquad Z_3 = 1.0\,\Omega,$$
$$R_{Z1} = Z_3/Z_1 = 0.10, \qquad R_{Z2} = Z_3/Z_2 = 0.13.$$

The loss-tangent ($\tan\delta_d$) is about 2×10^{-4} at 1 GHz, and the measured unloaded-Q value of the experimental DC-SIR was approximately 300 at 400 MHz. This result is approximately 90% of the calculated value obtained from Fig. 3.33.

Based on this resonator, we furthermore designed a 400 MHz-band Rx-filter for mobile communication applications. Specifications of the filter are as follows.

Center frequency	: f_0	$= 430\,\text{MHz}$
Pass-band bandwidth	: W	$> 10\,\text{MHz}$
Pass-band insertion loss	: L_0	$< 3.5\,\text{dB}$
Pass-band VSWR	:	< 1.5
Attenuation	: L_S	$> 50\,\text{dB}$ (at 400 MHz)

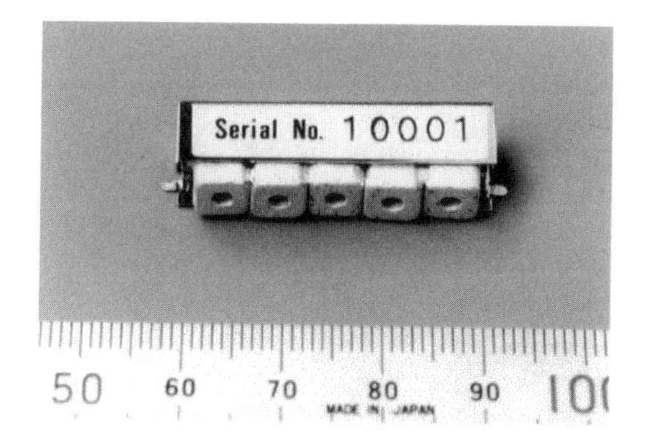

Fig. 3.35. Photograph of the experimental BPF

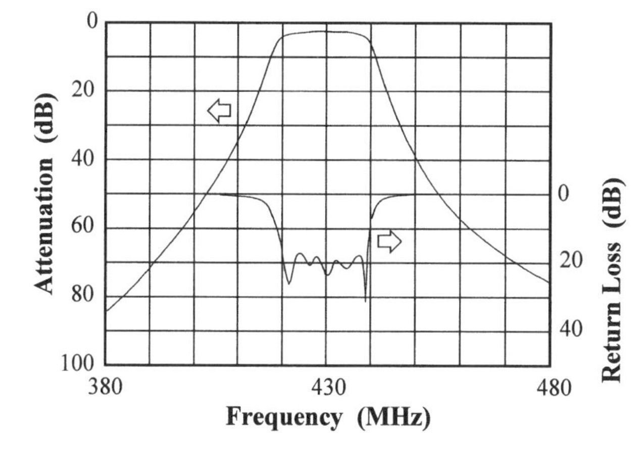

Fig. 3.36. Measured transmission responses of the experimental BPF

Applying the design method described in Sect. 3.2, a Chebyshev-type BPF with n=5 and pass-band ripple R=0.06 dB was designed in accordance with the above specifications. Figures 3.35 and 3.36 show a photograph and the measured transmission responses of the experimental filter. The physical dimensions are 32 mm(length)×6.5 mm(height)×9 mm(width) with a total volume of about 1.9 cm^{-3}, which enables a reduction of about 50% as compared to a conventional filter using helical resonators. Pass-band insertion losses measured 2.5 dB at midband and 3.3 dB at the band-edge, while an attenuation of more than 50 dB at 60 MHz off center frequency was achieved. Measured performance was in close agreement with design, satisfying requirements of a BPF for a 400 MHz band radio equipment while contributing to miniaturization of the radio terminal. As previously discussed, the DC-SIR structure enables applications to lower frequency bands where conventional helical resonators are otherwise used, and thus are expected to expand the applicable frequency range of dielectric coaxial resonators.

3.4.5 Dielectric Monoblock SIR-BPF

Dielectric ceramic-based resonators have an attractive feature which allows monolithic manufacturing of various structures, thus providing a design flexibility which enables complicated structures such as the DC-SIR. Applying this manufacturing strategy, furthermore, enables monolithic processing of a total BPF composed of several resonators of uniform structure. Such filters are fabricated in a single molding and sintering process, and are presently in practical use as dielectric monoblock filters.

Capacitive coupling between resonators often proves insufficient when designing a monoblock BPF, and thus it is often more practical to employ an electromagnetic coupling where the resonators are aligned and coupling is

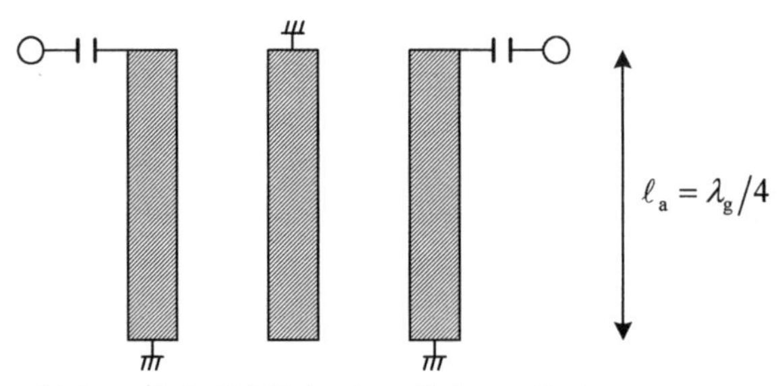

(a) Interdigital BPF (antiparallel coupling)

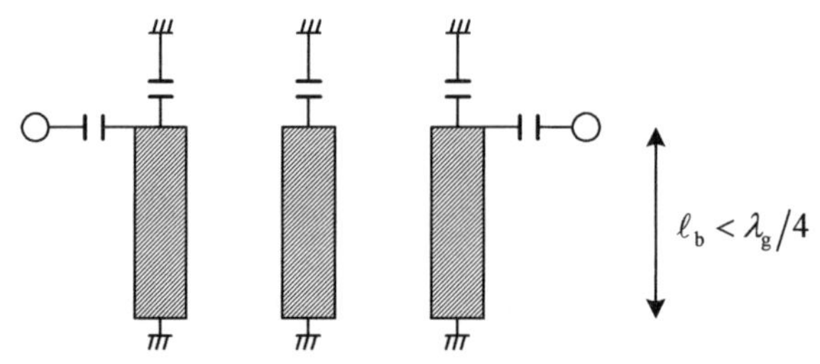

(b) Comb-line BPF (parallel coupling)

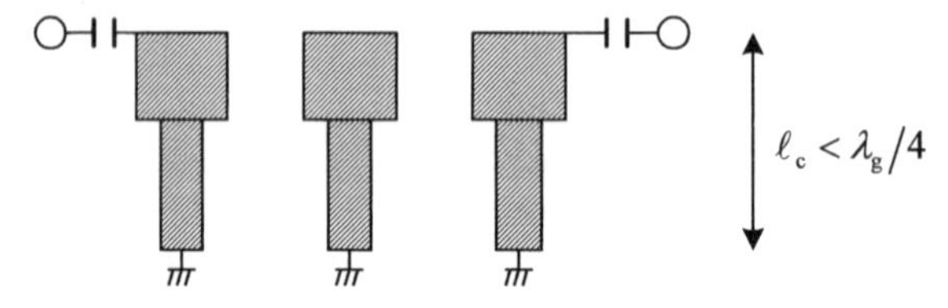

(c) Comb-line BPF using SIRs (parallel coupling)

Fig. 3.37. Basic resonator alignments for monoblock dielectric BPF. (**a**) Interdigital type. (**b**) Comb-line type. (**c**) Comb-line type using SIRs

controlled by spacing. Figure 3.37a illustrates the case of UIR, where an antiparallel configuration is required. Here also, the coupling strength between the resonators is determined by resonator spacing. The input and output couplings are realized by capacitive couplings in this example, but an inductive coupling structure, such as a tapping method, can equally be applied. Although this design possesses an extremely simple structure, it has the drawbacks of large size and heavy weight due to a resonator length of $\lambda_g/4$ and to the fact that the resonator spacing for a narrow band filter becomes exceedingly wide in order to reduce the strong coupling between the resonators.

A comb-line structure based on the parallel alignment of resonators can be applied when the coupling length and/or resonator length becomes shorter than $\lambda_g/4$. The unit resonator length of a comb-line BPF can and usually is shortened by loading a lumped-element capacitor at the open-end, as shown in Fig. 3.37b, but from this structure problems of increased circuit losses and stray components emerge. To overcome these disadvantages, Fig. 3.37c proposes a monoblock filter structure applying SIR [15]. Although the reduction of resonator length described in Sects. 3.4.2 and 3.4.4 cannot be expected from this filter structure, it does possess a practical advantage of cost-efficiency based on the above-mentioned monolithic manufacturing strategy.

Figure 3.38 shows a structural example of a three-stage dielectric monoblock filter using SIR. The whole surface of a monoblock ceramic shell is metalized after molding and sintering, followed by the removal of excess metal on the open-circuited end of the filter. Finally, the input and output electrodes are formed by thick-film printing technology. Frequency adjustment is achieved by trimming the metalized film at the open-ends of each resonator. This compact monoblock filter structure realizes good stability against environmental changes including temperature, humidity and vibration. These characteristics are eminently suitable for miniaturization of mobile radio communication equipment, and practical application of this structure has already been tested.

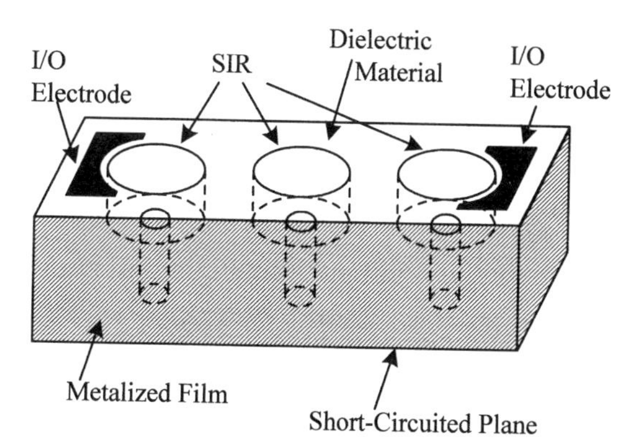

Fig. 3.38. Structural example of a dielectric monoblock SIR-BPF

3.5 Stripline SIR

3.5.1 Basic Structures and Features

The fundamental structure of a SIR can be defined as a TEM mode based transmission line possessing a change in impedance, and there exists no restriction on the physical structure of the TEM mode transmission line. Though our discussions have been focused mainly on coaxial transmission lines, stripline, micro-stripline and coplanar-line structures formed on dielectric substrates can also be applied as transmission lines of SIR. Yet to fully appreciate the features of SIR, we need to keep in mind that it is desirable to adopt a structure allowing a large impedance ratio R_Z, and from this point of view we exclude the coplanar transmission line structure due to its narrow design range of impedance. The basic structure of a SIR using a micro-stripline configuration is shown in Fig. 3.39. A stripline SIR possesses a tri-plate structure consisting of an extra dielectric substrate and metalized ground plane covering the resonator center conductor, but the basic resonator structure is equivalent to that of a micro-stripline.

Ground via-holes as shown in Fig. 3.39 are essential for $\lambda_g/4$ type SIR of a stripline configuration. This necessity raises inevitable problems such as increased losses and resonance frequency shifts caused by the parasitic components generated near via-holes. For this reason the application of $\lambda_g/4$ type SIR is limited to filters of special use where, for example, miniaturization is prior to low insertion losses. Generally speaking, open-ended $\lambda_g/2$ type SIR possess a far wider applicable range, thus are more available for RF and microwave circuits.

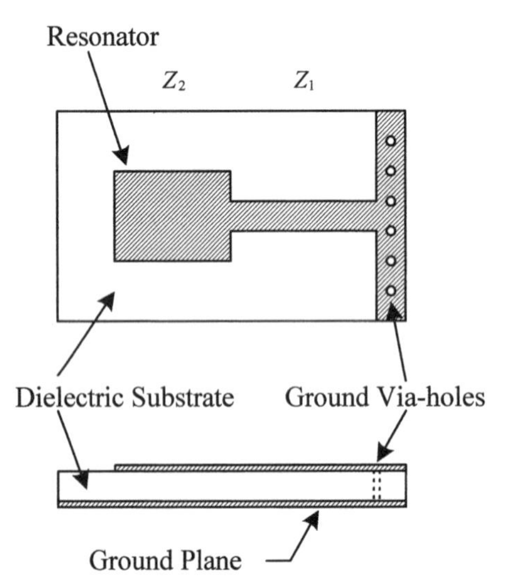

Fig. 3.39. Basic structure of a micro-stripline SIR

Since stripline and micro-stripline resonators are formed on dielectric substrate, the use of a dielectric material with high permittivity proves to be most effective for miniaturization of filters. A stripline resonator, which has a tri-plate structure, possesses the same wavelength reduction factor as a coaxial resonator, whereas in the case of micro-stripline the reduction factor decreases due to the inhomogeneous dielectric medium.

The unloaded-Q of stripline and micro-stripline resonators is dependent on the line width and the substrate thickness, and it becomes difficult to design high Q value resonator as compared to a coaxial type structure. This is because the center conductor of a coaxial resonator possesses a large surface area along with a uniform current distribution, whereas in the case of a stripline structure, ohmic losses are apt to increase due to the current concentration at the edges of the stripline center conductor. However, the stripline SIR has a distinct feature that allows for cost-efficient fabrication of various complicated structures due to a manufacturing process based on thick film and/or thin film processing technology. Thus, it can be concluded that stripline SIR are an available resource for filters which call for small size rather than low losses, and for microwave circuits requiring integration to active devices.

3.5.2 Coupling Between Resonators

It is a well-known method to apply parallel coupled-lines to obtain interstage coupling between resonators. In the case of SIR, the coupling circuit is electrically expressed as two pairs of coupled-lines as illustrated in Fig. 3.40. This circuit differs from that of the UIR. Coupled-lines 1 including a short-circuited section can be analyzed by even- and odd-mode impedance Z_{0e1}, Z_{0o2} and coupled line length θ_1. Inductive coupling is dominant in this por-

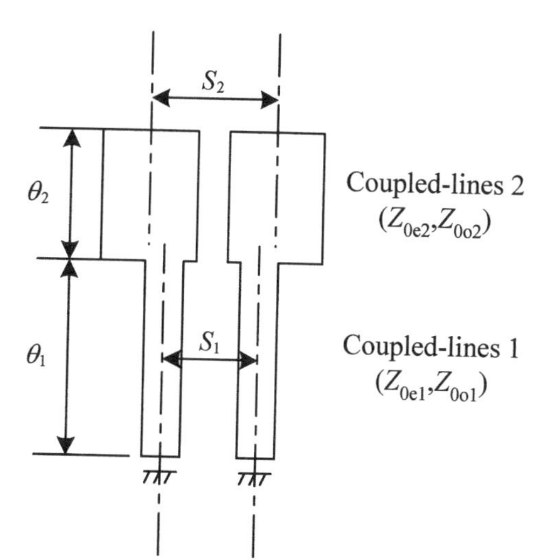

Fig. 3.40. Interstage coupling structure of stripline SIRs

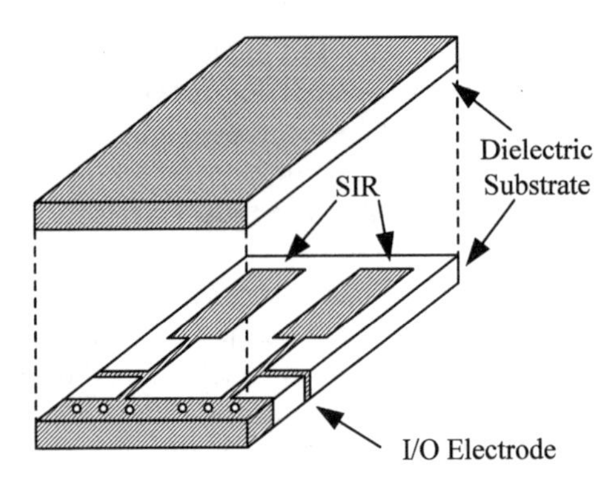

Fig. 3.41. Structural example of a stripline BPF

tion due to a large current flow near the short-circuited point. Coupled-lines 2 includes the open-end of the resonator, its electrical parameters being defined as Z_{0e2}, Z_{0o2}, and θ_2. Capacitive coupling is dominant in this portion due to a high voltage near the open-end. Z_1 and Z_2 for single transmission line SIR are given as the geometric means of even and odd-mode impedance as

$$Z_1 = \sqrt{Z_{0e1} \cdot Z_{0o1}},$$
$$Z_2 = \sqrt{Z_{0o2} \cdot Z_{0o2}}.$$

Thus, the even- and odd-mode impedance cannot be determined independently. Two pairs of coupled lines enable a more flexible design, while on the other hand this becomes a disadvantage because the coupling circuits cannot be determined uniquely. Although theoretical analysis of these two pairs of coupled-lines has been reported, a more generalized approach using a conventional microwave circuit simulator is also accepted, which we describe in the appendix.

3.5.3 Stripline SIR-BPF

From a practical point of view, a stripline SIR-BPF is not suitable for filters which require low-loss and steep cut-off characteristics, and is often applied to meet with demands which list miniaturization as the top priority [16,17]. A practical example actually in use is a two-stage BPF connecting the RF amplifier and mixer of a mobile communication terminal. Figure 3.41 shows a structural example of this two-stage BPF using stripline SIR. By employing tapping couplings for the input and output terminals and thus eliminating any additional components, this filter structure can be realized genuinely by thick film printing technology.

An example of typical filter transmission characteristics is shown in Fig. 3.42. The figure suggests that this filter features a steep attenuation characteristic due to an attenuation pole generated at the lower side of the pass-band. The existence of the two pairs of coupled-lines enables this design,

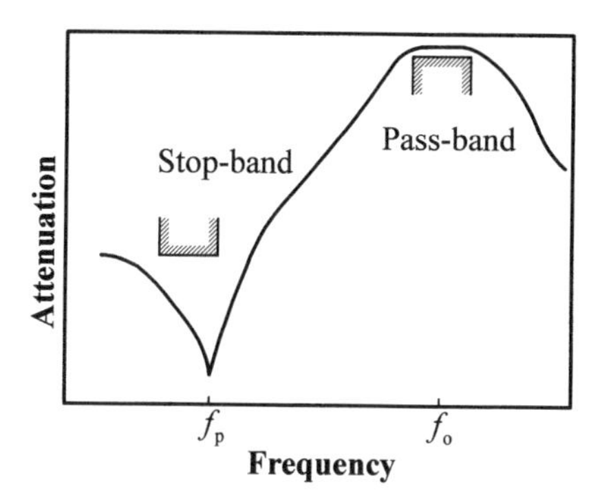

Fig. 3.42. Example of typical frequency response of a stripline SIR-BPF

where the attenuation pole frequency is controlled by the coupling strength of these coupled-lines. A contrary design where the attenuation pole is generated at the upper side of the pass-band can also be realized by choosing the proper combination of these couplings. Analysis of such attenuation pole design will be discussed in Chap. 5.

4. Half-Wavelength-Type SIR

The basic structure, electrical characteristics, and application examples of a $\lambda_g/2$ type SIR are discussed in this chapter. Although the $\lambda_g/4$ type SIR is proven to be the most suitable structure for miniaturization, in practical application the $\lambda_g/2$ type SIR realizes far more RF devices than does the $\lambda_g/4$ type SIR. This is due to the fact that the $\lambda_g/2$ type SIR is usually composed of a stripline configuration, thus allowing a wide range of geometrical structures while possessing good affinity with active devices.

Figure 4.1 illustrates typical structural variations of the $\lambda_g/2$ type SIR. Structures (a), (b), and (c) in the figure are equivalent from a circuit topology point of view, though the geometrical structures vary between linear, U (hairpin), and ring-shape. The resonator shown in (d) has a U-shaped structure resembling resonator (b), but also possesses inner coupled-lines utilizing both open-ends for miniaturization. Structure (e) is an improved structure of (d) where the space factor is remarkably extended for further miniaturization. The figure demonstrates the expanded flexibility of the $\lambda_g/2$ type SIR for circuit pattern layout and coupling circuit integration.

Features of the $\lambda_g/2$ type SIR are summarized as follows.

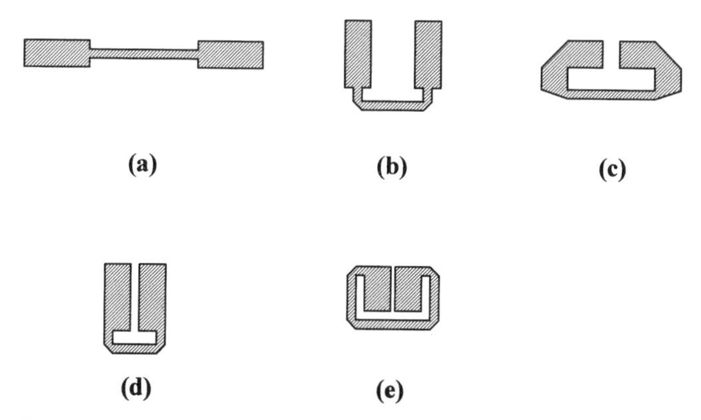

Fig. 4.1. Some structural variations of $\lambda_g/2$ type SIR. (**a**) Straight type. (**b**) Hairpin type. (**c**) Ring type. (**d**) Hairpin type with internal coupling. (**e**) Ring type with internal coupling

(i) A highly suitable structure for hybrid ICs due to a stripline configuration without via-holes,
(ii) A variety of geometrical structures with equivalent circuit topology based on the availability of fabrication by thick-film printing or thin-film photo-etching technology,
(iii) Good affinity with active devices and practical applicability to oscillators and mixers,
(iv) A wide range of applicable frequency from RF to microwave by proper selection of substrate material.

In this chapter, the basic characteristics of a $\lambda_g/2$ type SIR based on straight transmission lines are presented. This is followed by discussions on the inverter expression of coupling circuits composed of parallel coupled-lines, which becomes an important issue for practical applications. Based on these results, the generalized design method of a parallel-coupled SIR-BPF with arbitrary coupling length is derived, and two design examples are demonstrated. Next, analysis and basic characteristics of a SIR possessing internal coupling between the two open-ends are described, followed by design examples of miniaturized filters including superconductor filters. Finally, active circuit applications are illustrated through examples of balanced mixers and push-push oscillators utilizing the features of SIR, and case studies on actual R&D issues are presented.

4.1 Stripline $\lambda_g/2$ Type SIR

4.1.1 Basic Characteristics

The basic structure of a $\lambda_g/2$ type SIR is shown in Fig. 4.2. Analysis of this resonator is performed in the same manner as a $\lambda_g/4$ type SIR discussed in Chap. 2. Thus, for a direct analysis, input admittance Y_i seen from an open-end must be obtained, in this case given as [1],

$$Y_\mathrm{i} = \mathrm{j}Y_2 \cdot \frac{2(R_Z \tan\theta_1 + \tan\theta_2)(R_Z - \tan\theta_1 \tan\theta_2)}{R_Z(1-\tan^2\theta_1)(1-\tan^2\theta_2) - 2(1-R_Z^2)\tan\theta_1\tan\theta_2}. \tag{4.1}$$

Resonance conditions are obtained by taking $Y_\mathrm{i} = 0$, thus giving

$$R_Z = \frac{Z_2}{Z_1} = \tan\theta_1 \tan\theta_2.$$

As previously described, this equation is common for all SIR structures. Consequently, resonator electrical length, spurious frequencies, and susceptance slope parameters can be discussed in the same manner using (4.1). In the case of $\theta_1 = \theta_2 = \theta$, (4.1) is simplified as

$$Y_\mathrm{i} = \mathrm{j}Y_2 \frac{2(1+R_Z)(R_Z - \tan^2\theta)\tan\theta}{R_Z - 2(1+R_Z+R_Z^2)\tan\theta}. \tag{4.2}$$

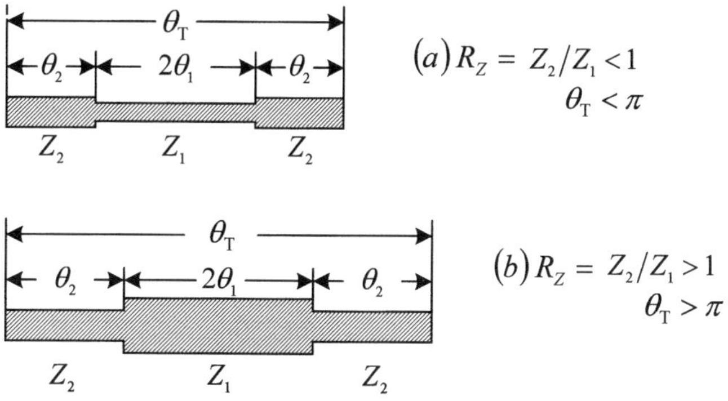

$(a)\ R_Z = Z_2/Z_1 < 1$, $\theta_T < \pi$

$(b)\ R_Z = Z_2/Z_1 > 1$, $\theta_T > \pi$

Fig. 4.2. Basic structure of a stripline $\lambda_g/2$ type SIR. (**a**) $R_Z < 1$. (**b**) $R_Z > 1$

Thus, resonance conditions are expressed as

$$\theta = \theta_0 = \tan^{-1}\sqrt{R_Z}. \tag{4.3}$$

Spurious responses of a $\lambda_g/2$ type SIR become more critical compared with $\lambda_g/4$ type SIR [1]. This requires the designer to consider the spurious responses of higher resonance modes which otherwise would be neglected for $\lambda_g/4$ type SIR. By expressing the spurious resonance frequencies as $f_{SB1}, f_{SB2}, f_{SB3}$ and the corresponding θ as $\theta_{S1}, \theta_{S2}, \theta_{S3}$, we obtain from (4.2)

$$\begin{aligned}
\theta_{S1} &= \pi/2, \\
\theta_{S2} &= \tan^{-1}(-\sqrt{R_Z}) = \pi - \theta_0, \\
\theta_{S3} &= \pi.
\end{aligned}$$

Thus,

$$\frac{f_{SB1}}{f_0} = \frac{\theta_{S1}}{\theta_0} = \frac{\pi}{2\tan^{-1}\sqrt{R_Z}}, \tag{4.4a}$$

$$\frac{f_{SB2}}{f_0} = \frac{\theta_{S2}}{\theta_0} = 2\left(\frac{f_{SB1}}{f_0}\right) - 1, \tag{4.4b}$$

$$\frac{f_{SB3}}{f_0} = \frac{\theta_{S3}}{\theta_0} = 2\left(\frac{f_{SB1}}{f_0}\right). \tag{4.4c}$$

4.1.2 Equivalent Expressions for Parallel Coupled-Lines Using Inverter

A conventional BPF suitable for MIC utilizes parallel-coupled open-ended $\lambda_g/2$ UIR, where the coupling length is generally chosen as $\lambda_g/4$ (equivalent to a coupling angle of $\pi/4$ radian), making it possible to derive the design formula under this condition. When applying SIR as a unit resonator element, the input/output coupling and interstage coupling between the resonators can still be realized by the same parallel coupled-lines, yet the coupling angle is

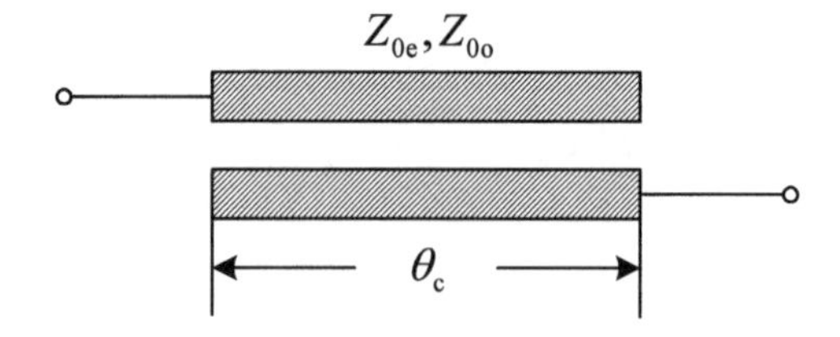

Fig. 4.3. Parallel coupled striplines and their electrical parameters

not limited to $\pi/4$ radian. This is because in the case of SIR, the parallel coupled-lines are realized from the Z_2 section including the open-end of the SIR as illustrated in Fig. 4.2. Thus, when adopting SIR, the designer must consider the coupling angle corresponding to an arbitrary line length of the Z_2 section based on the resonator structure [1].

Figure 4.3 shows the distributed coupling circuit applied to a stripline parallel-coupled BPF. The electrical parameters of this circuit are expressed by even- and odd-mode impedance Z_{0e}, Z_{0o}, and electrical coupling angle θ. In the case of a tri-plate stripline structure, θ for both even- and odd-modes are the same because the phase velocity of both modes is equal. On the contrary, for a micro-stripline configuration, the phase velocity differs for both modes, thus making it necessary to distinguish the coupling angles as θ_{0e} and θ_{0o} to obtain a strict analysis. However, this difference has little influence on the coupling strength, and thus here we consider $\theta_{0e} = \theta_{0o} = \theta_c$ for simplicity. Under this condition, the impedance matrix $[Z]$ of the parallel coupled-lines is given as

$$[Z] = \begin{bmatrix} Z_{11} & Z_{12} \\ Z_{21} & Z_{22} \end{bmatrix} = \begin{bmatrix} -\mathrm{j}\dfrac{Z_{0e}+Z_{0o}}{2}\cot\theta_c & -\mathrm{j}\dfrac{Z_{0e}-Z_{0o}}{2}\operatorname{cosec}\theta_c \\ -\mathrm{j}\dfrac{Z_{0e}-Z_{0o}}{2}\operatorname{cosec}\theta_c & -\mathrm{j}\dfrac{Z_{0e}+Z_{0o}}{2}\cot\theta_c \end{bmatrix}. \tag{4.5}$$

Since a BPF is generally expressed as a cascaded connection of resonators and coupling circuits, the above impedance matrix $[Z]$ is transformed into the fundamental matrix $[F_a]$ in an attempt to simplify mathematical operations.

$$[F_a] = \begin{bmatrix} A & B \\ C & D \end{bmatrix} \tag{4.6}$$

$$A = D = \frac{Z_{11}}{Z_{21}} = \frac{Z_{0e}+Z_{0o}}{Z_{0e}-Z_{0o}}\cos\theta_c,$$

$$B = \frac{Z_{11}Z_{22}-Z_{12}Z_{21}}{Z_{21}} = \mathrm{j}\frac{(Z_{0e}-Z_{0o})^2+(Z_{0e}+Z_{0o})^2\cos^2\theta_c}{2(Z_{0e}-Z_{0o})\sin\theta_c},$$

$$C = \frac{1}{Z_{21}} = \mathrm{j}\frac{2\sin\theta_c}{Z_{0e}-Z_{0o}}.$$

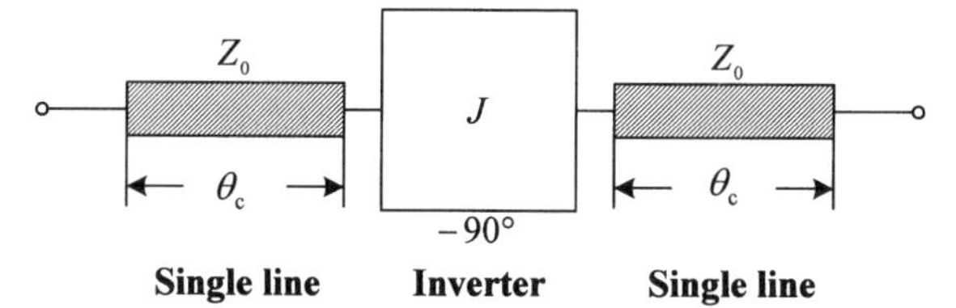

Fig. 4.4. An equivalent circuit of parallel coupled-lines using a J-Inverter

Although the circuit matrix for parallel coupled-lines can be expressed as (4.5) or (4.6), we find it extremely difficult to directly synthesize the BPF from these equations. To overcome this problem, we introduce an equivalent circuit composed of two single lines and an admittance inverter, which we treated as an ideal coupling circuit without frequency dependency, as shown in Fig. 4.4.

An admittance inverter is defined as a two- port passive circuit of which the input admittance is inversely proportional to the load admittance Y_L [2]. Taking the admittance inverter parameter as J, the fundamental matrix of an admittance inverter is expressed as

$$[F] = \begin{bmatrix} 0 & \pm j\dfrac{1}{J} \\ \pm jJ & 0 \end{bmatrix}. \tag{4.7}$$

Assuming the load admittance of this circuit as Y_L, Y_i is given as

$$Y_i = \frac{C + DY_L}{A + BY_L}.$$

Substituting the matrix elements from (4.7), we obtain

$$Y_i = \frac{C}{BY_L} = \frac{J^2}{Y_L}.$$

It can be seen from the above equation that Y_i is inversely proportional to Y_L. A wellknown example of an admittance inverter is a quarter-wavelength transmission line. Assuming a loss-free transmission line with impedance and line length expressed as Z_0 and θ, the fundamental matrix is given as

$$[F] = \begin{bmatrix} \cos\theta & jZ_0 \sin\theta \\ j\dfrac{\sin\theta}{Z_0} & \cos\theta \end{bmatrix}. \tag{4.8}$$

Considering $\theta = \pi/2$, we obtain

$$[F] = \begin{bmatrix} 0 & jZ_0 \\ j\dfrac{1}{Z_0} & 0 \end{bmatrix}.$$

The above equation satisfies the condition of an admittance inverter as defined in (4.7), and the admittance inverter parameter is given as

$$J = \frac{1}{Z_0}.$$

Next, returning to Fig. 4.4 we derive the total fundamental matrix $[F_b]$. Referring to (4.7) and (4.8), $[F_b]$ is given as

$$[F_b] = \begin{bmatrix} \cos\theta_c & jZ_0\sin\theta_c \\ \dfrac{j\sin\theta_c}{Z_0} & \cos\theta_c \end{bmatrix} \begin{bmatrix} 0 & \dfrac{j}{J} \\ -jJ & 0 \end{bmatrix} \begin{bmatrix} \cos\theta_c & jZ_0\sin\theta_c \\ \dfrac{j\sin\theta_c}{Z_0} & \cos\theta_c \end{bmatrix}$$

$$= \begin{bmatrix} \left(JZ_0 + \dfrac{1}{JZ_0}\right)\sin\theta_c\cos\theta_c & j\left(JZ_0^2\sin\theta_c - \dfrac{\cos^2\theta_c}{J}\right) \\ j\left(\dfrac{\sin^2\theta_c}{JZ_0^2} - J\cos^2\theta_c\right) & \left(JZ_0 + \dfrac{1}{JZ_0}\right)\sin\theta_c\cos\theta_c \end{bmatrix}. \tag{4.9}$$

Since $[F_a] = [F_b]$, by equalizing each corresponding matrix element we obtain

$$\frac{Z_{0e} + Z_{0o}}{Z_{0e} - Z_{0o}}\cos\theta_c = \left(JZ_0 + \frac{1}{JZ_0}\right)\sin\theta_c\cos\theta_c, \tag{4.10}$$

$$\frac{(Z_{0e} - Z_{0o})^2 - (Z_{0e} + Z_{0o})^2\cos^2\theta_c}{2(Z_{0e} - Z_{0o})\sin\theta_c} = JZ_0^2\sin^2\theta_c - \frac{\cos^2\theta_c}{J}, \tag{4.11}$$

$$\frac{2\sin\theta_c}{Z_{0e} - Z_{0o}} = \frac{\sin^2\theta_c}{JZ_0^2} - J\cos^2\theta_c. \tag{4.12}$$

The above simultaneous equations are not independent of each other, and any two equations among the three are valid for solution. Solving (4.10) and (4.11) with respect to Z_{0e} and Z_{0o},

$$\frac{Z_{0e}}{Z_0} = \frac{1 + JZ_0\mathrm{cosec}\theta_c + J^2Z_0^2}{1 - J^2Z_0^2\cot^2\theta_c}$$
$$= \frac{1 + (J/Y_0)\mathrm{cosec}\theta_c + (J/Y_0)^2}{1 - (J/Y_0^2)\cot^2\theta_c}, \tag{4.13}$$

$$\frac{Z_{0o}}{Z_0} = \frac{1 - JZ_0\mathrm{cosec}\theta_c + J^2Z_0^2}{1 - J^2Z_0^2\cot^2\theta_c}$$
$$= \frac{1 - (J/Y_0)\mathrm{cosec}\theta_c + (J/Y_0)^2}{1 - (J/Y_0)^2\cot^2\theta_c}. \tag{4.14}$$

These equations are generalized expressions for parallel coupled-lines with arbitrary coupling length. For the case of a quarter-wavelength coupling, substituting $\theta_c = \pi/2$ into (4.13) and (4.14) gives

$$\frac{Z_{0e}}{Z_0} = 1 + \frac{J}{Y_0} + \left(\frac{J}{Y_0}\right)^2, \tag{4.15}$$

$$\frac{Z_{0o}}{Z_0} = 1 - \frac{J}{Y_0} + \left(\frac{J}{Y_0}\right)^2. \tag{4.16}$$

These equations correspond to Cohn's results [3]. We understand from the above discussion that generalized circuit parameters of parallel coupled-lines with arbitrary coupling length are determined by specifying inverter parameters required for the filter to be designed.

4.1.3 Synthesis of Stripline Parallel-Coupled SIR-BPF

The basic configuration of an n-stage BPF under consideration is shown in Fig. 4.5. SIR composing this filter can be of a uniform structure or a combination of different structures. The synthesis method discussed here provides a generalized design technique for parallel coupled resonator filters, and the design method of a conventional parallel-coupled UIR-BPF can be derived from this method as a special case where $R_Z = 1$ and $\theta_C = \pi/2$.

The following design conditions are introduced for simplicity and easy understanding.

(i) SIR structure:
Electrical line length is fixed at $\theta_1 = \theta_2 = \theta$, and the impedance ratio and corresponding admittance slope parameters of the j-th resonator are defined as R_{Zj} and b_j.

(ii) Coupling conditions:
Input/output coupling and interstage coupling is achieved by parallel coupled-lines realized by the Z_2 line section including the open-ends.

(iii) Input and output impedance:
Both input and output port impedance is Z_0, and the characteristic impedance Z_2 of the unit SIR is also chosen as Z_0 for design simplicity.

These conditions are set for the practical convenience of simplifying design equations and procedures, and are not essential to the analysis.

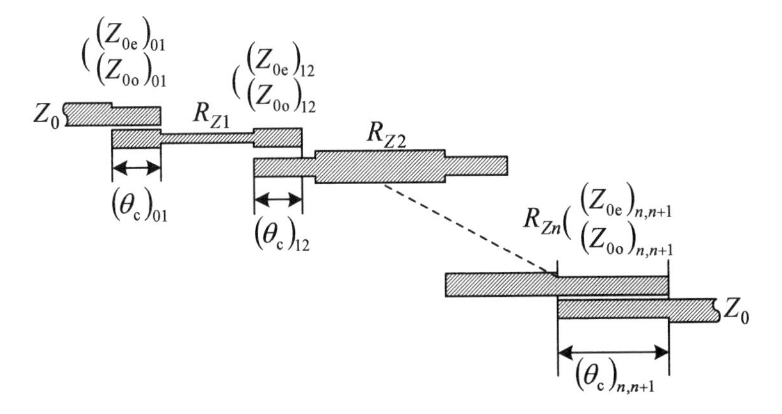

Fig. 4.5. Generalized configuration of n-stage BPF using stripline SIRs

When designing a coupled-line one should keep in mind that the coupling line length should be kept as long as possible while obeying the structural restrictions of the SIR, thus securing a wide line spacing between coupled-lines. Especially for the first and final stages of a BPF where the required coupling strength tends to be a decade higher than interstage couplings, the application of UIR possessing a coupling length of $\pi/2$ proves to be an effective solution for practical circuit patterning.

Basic design procedures for stripline SIR-BPF resemble those of coaxial SIR-BPF discussed in Chap. 3. The actual procedures are shown as follows.

(a) Calculation of BPF Basic Parameters

According to the given specifications (center frequency, bandwidth, attenuation, etc.), the basic parameters of the BPF such as number of stages, relative bandwidth, and element values are determined after specifying the filter response type such as Chebyshev or Butterworth.

(b) Selection of SIR Basic Structure

When designing the unit SIR, the impedance ratio R_Z must first be specified in accordance with the target application of the BPF. For example, $R_Z < 1$ is recommended for a design where miniaturization is required, while $R_Z > 1$ is desirable for a contrary design emphasizing low-loss characteristics. Furthermore, the R_Z value of all unit SIR can be designed to be equivalent, or a combination of different values can be chosen, which is known to be very effective for expanding the stop-band of the BPF.

Although the fundamental design of SIR-BPF can be achieved by the method described in Sect. 4.1, computer simulations presented in the appendix can be applied when rigorous estimations of unloaded-Q, coupling factor, and resonance frequency including junction and fringing effects [4,5] are required.

(c) Calculation of Inverter Parameters

Once the fundamental SIR structure is determined, the sucseptance slope parameter b_j can be obtained from (2.19) and (2.22) or (4.1). Since relative bandwidth w and element value g_j are already given, the admittance inverter parameters $J_{j,j+1}$ can be calculated as follows:

$$J_{01} = \sqrt{\frac{Y_0 b_1 w}{g_0 g_1}}, \tag{4.17a}$$

$$J_{j,j+1} = w\sqrt{\frac{b_j b_{j+1}}{g_j g_{j+1}}} \qquad (j = 1 \text{ to } n-1), \tag{4.17b}$$

$$J_{n,n+1} = \sqrt{\frac{Y_0 b_n w}{g_n g_{n+1}}}. \tag{4.17c}$$

(d) Determination of Coupled-Lines Parameters and Structure

With the admittance inverter parameters determined, the coupled-lines electrical parameters Z_{0e} and Z_{0o} can be obtained from (4.10) and (4.11) by specifying coupling line length or coupling angle θ_C. Furthermore, with Z_{0e} and Z_{0o} determined, line width W and line spacing S of the coupled-lines is obtained in accordance with the transmission line structure such as tri-plate stripline, micro-stripline, etc. Synthesis of these coupled-lines can be achieved by using a commercially available general-purpose microwave circuit simulator. A tri-plate stripline structure shown in Fig. 4.6 is adopted for the design example later described. In this case, line width W and line spacing S can be analytically derived as follows.

As shown in the figure, the dielectric constant of the substrate material, substrate thickness, line width, and line spacing of the coupled-lines are expressed as ε_r, H, W_c and S, respectively, while the line width of a single line is expressed as W_s. When the center conductor thickness t is extremely small compared to the substrate thickness $H(t << H)$, the single line characteristic impedance Z_0 and coupled-lines even- and odd-mode impedances Z_{0e} and Z_{0o} are obtained as follows.

Single line:

$$Z_0 = \frac{30\pi}{\sqrt{\varepsilon_r}} \cdot \frac{1}{\frac{W_s}{H} + \frac{2}{\pi}\ln 2}. \tag{4.18}$$

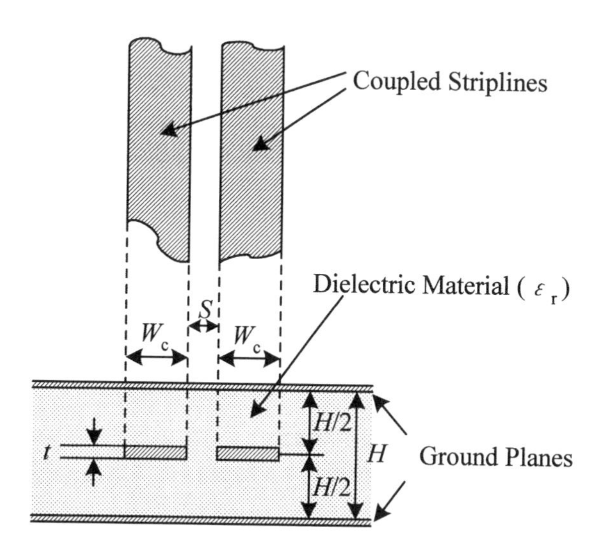

Fig. 4.6. Physical parameters of tri-plate coupled striplines

Coupled-lines:

$$Z_{0\mathrm{e}} = \frac{30\pi}{\sqrt{\varepsilon_\mathrm{r}}} \cdot \frac{1}{\frac{W_s}{H} + \frac{1}{\pi}\ln 2 + \frac{1}{\pi}\ln\left\{1 + \tan\mathrm{h}\left(\frac{\pi S}{2H}\right)\right\}}. \tag{4.19a}$$

$$Z_{0\mathrm{o}} = \frac{30\pi}{\sqrt{\varepsilon_\mathrm{r}}} \cdot \frac{1}{\frac{W_s}{H} + \frac{1}{\pi}\ln 2 + \frac{1}{\pi}\ln\left\{1 + \cot h\left(\frac{\pi S}{2H}\right)\right\}}. \tag{4.19b}$$

When performing transmission line synthesis, it is required to obtain the structural parameters for a given line impedance from the above equations. For a single transmission line, (4.18) can be solved directly, giving,

$$\frac{W_\mathrm{s}}{H} = \frac{30\pi}{Z_0\sqrt{\varepsilon_\mathrm{r}}} - \frac{2}{\pi}\ln 2. \tag{4.20}$$

As for coupled-lines, (4.19a) and (4.19b) can be solved with respect to W_c and S by introducing parameters A and B as follows.

$$\frac{W_\mathrm{c}}{H} = A - \frac{1}{\pi}\ln\{1 + \exp[\pi(A - B)]\},$$
$$\frac{S}{H} = \frac{1}{\pi}\ln\left\{\frac{1 + \exp[\pi(A - B)]}{1 - \exp[\pi(A - B)]}\right\}, \tag{4.21}$$

where

$$A = \frac{30\pi}{Z_{0\mathrm{e}}\sqrt{\varepsilon_\mathrm{r}}} - \frac{1}{\pi}\ln 2,$$
$$B = \frac{30\pi}{Z_{0\mathrm{o}}\sqrt{\varepsilon_\mathrm{r}}} - \frac{1}{\pi}\ln 2.$$

The above discussion suggests that it is possible to determine physical dimensions of a transmission line by giving Z_0 for single line and $Z_{0\mathrm{e}}$ and $Z_{0\mathrm{o}}$ for coupled-lines.

4.1.4 Filter Design Examples

The validity of the parallel coupled SIR-BPF design method presented in the previous subsection is demonstrated through the following examples of experimental filters actually designed and fabricated [1]. We present two BPF examples, one composed of SIR of uniform structure, and another applying a combination of different SIR structures to improve stop-band characteristics.

(a) BPF Applying Uniform SIR Structure

The following are design specifications of the experimental uniform structure BPF.

Center frequency	: $f_0 = 1000\,\mathrm{MHz}$
Pass-band width	: $W > 20\,\mathrm{MHz}$
Attenuation	: $L_\mathrm{S} > 45\,\mathrm{dB}$ at $f_0 \pm 100\,\mathrm{MHz}$
Spurious resonance frequency	: $f_\mathrm{S} > 2.5 f_0$

Based on the method described in Sect. 3.2.2, we first determine the filter fundamental parameters which meet the above specifications.

Filter response type : Chebyschev
Number of stages : $n = 4$
Pass-band ripple : $R = 0.01\,\text{dB}$
Relative bandwidth : $w = 0.04$

Thus, element values g_j are obtained as,

$$g_0 = 1.000 \qquad g_1 = 0.7129$$
$$g_2 = 1.2004 \qquad g_3 = 1.3213$$
$$g_4 = 0.6476 \qquad g_5 = 1.1007.$$

Insertion loss at the center frequency can be expressed as

$$L_0 = \frac{4.434}{Q_0 w} \sum_{\mathrm{j}=1}^{4} g_{\mathrm{j}} = \frac{422}{Q_0} \text{ (dB)}.$$

With the fundamental parameters specified, we next determine the physical parameters of the experimental BPF. The resonator structure is first decided, which requires consideration of spurious frequency conditions specified as above $2.5f_0$. From the discussion in Sect. 4.1.1, the lowest spurious resonance frequency f_{SB1} of a $\lambda_g/2$ type SIR with line length $\theta_1 = \theta_2$ is given from (4.4a) as,

$$\frac{f_{SB1}}{f_0} = \frac{\pi}{2\tan^{-1}\sqrt{R_Z}}.$$

Thus, to satisfy $f_S = f_{SB1} > 2.5f_0$, conditions for R_Z are given as

$$\tan^{-1}\sqrt{R_Z} < \frac{\pi}{2} \cdot \frac{1}{2.50} = 0.6283\text{radians},$$
$$R_Z < (\tan 0.6283)^2 = 0.528.$$

Considering these results, we choose a R_Z value of 0.5.

The parallel coupled section is realized by the Z_2 section, where impedance Z_2 is assumed as $50\,\Omega$. Thus, we obtain

$$Z_1 = \frac{Z_2}{R_Z} = \frac{50}{0.50} = 100\,\Omega,$$
$$\theta_c = \theta_0 = \tan^{-1}\sqrt{R_Z} = 0.616\,\text{radians} = 35.3°,$$
$$b_0 = 2\theta_0 Y_2 = 0.0246\,\text{S}.$$

The lowest spurious resonance frequency f_S is calculated as

$$f_S = f_{SB1} = \frac{\pi}{2\theta_0} f_0 = 2.55f_0.$$

These results satisfy the requirements for spurious response. Other spurious responses can also be calculated from (4.4b) and (4.4c), the results are given below for reference.

$$f_{\mathrm{SB2}} = (2f_{\mathrm{SB1}}/f_0 - 1)f_0 = 4.10f_0,$$
$$f_{\mathrm{SB3}} = 2f_{\mathrm{SB1}} = 5.10f_0.$$

These results suggest that the spurious frequencies are shifted from the integer multiples of fundamental frequency f_0. This property can effectively be applied to the output filter of amplifiers and oscillators for the suppression of upper harmonic frequency components. The inverter parameters of the coupled section can be determined from (4.17a), (4.17b) and (4.17c) based on the values obtained of g_j, b, w, Y_0.

$$\frac{J_{01}}{Y_0} = \frac{J_{45}}{Y_0} = 0.2628,$$
$$\frac{J_{12}}{Y_0} = \frac{J_{34}}{Y_0} = 0.05323,$$
$$\frac{J_{23}}{Y_0} = 0.03910.$$

Considering the above results, the even- and odd-mode impedance of each coupled section can be determined by (4.13) and (4.14) as follows,

$$\begin{cases} (Z_{0\mathrm{e}})_{01} = (Z_{0\mathrm{e}})_{45} = 88.4\,\Omega \\ (Z_{0\mathrm{e}})_{01} = (Z_{0\mathrm{o}})_{45} = 35.6\,\Omega, \end{cases}$$

$$\begin{cases} (Z_{0\mathrm{e}})_{12} = (Z_{0\mathrm{e}})_{34} = 55.1\,\Omega \\ (Z_{0\mathrm{o}})_{12} = (Z_{0\mathrm{o}})_{34} = 45.8\,\Omega, \end{cases}$$

$$\begin{cases} (Z_{0\mathrm{e}})_{23} = 53.6\,\Omega \\ (Z_{0\mathrm{e}})_{01} = 46.8\,\Omega. \end{cases}$$

Having determined the electrical parameters, we next consider the physical dimensions of the SIR-BPF. Prior to this design process, required specifications of transmission line structure and substrate material are chosen as below.

Transmission line structure : Tri-plate stripline
Substrate material : Glass-fiber laminate
Relative dielectric constant $\varepsilon_r = 2.6$
Substrate height $H = 3.13\,\mathrm{mm}$
Center conductor thickness
$t = 0.018\,\mathrm{mm}$
Loss tangent $\tan\delta_{\mathrm{d}} \cong 1.0 \times 10^{-3}$.

Based on the above conditions, the physical dimensions of the resonator are first determined. As shown in Fig. 4.2, line lengths of section Z_1, Z_2 and total line length are expressed as $2\ell_1, 2\ell_2$, and ℓ_{T}, respectively, and thus ℓ_1 is given as,

$$\ell_1 = \frac{\theta_0}{2\pi} \cdot 300 \cdot \frac{1}{\sqrt{\varepsilon_{\mathrm{r}}}} = 18.22\,\mathrm{mm}.$$

Table 4.1. Physical dimensions of transmission line for SIRs ($H = 3.13$mm, $\varepsilon_r = 2.60$)

Characteristic Impedance (Ω)		Line Width (mm)	Line Spacing (mm)
Single Line	$Z_1 = 100$	0.45	–
	$Z_2 = Z_0 = 50$	2.27	–
Coupled-Lines	$Z_{0e} = 88.43$, $Z_{0o} = 35.61$	1.33	0.09
	$Z_{0e} = 55.06$, $Z_{0o} = 45.79$	2.20	1.12
	$Z_{0e} = 53.63$, $Z_{0o} = 48.83$	2.24	1.41

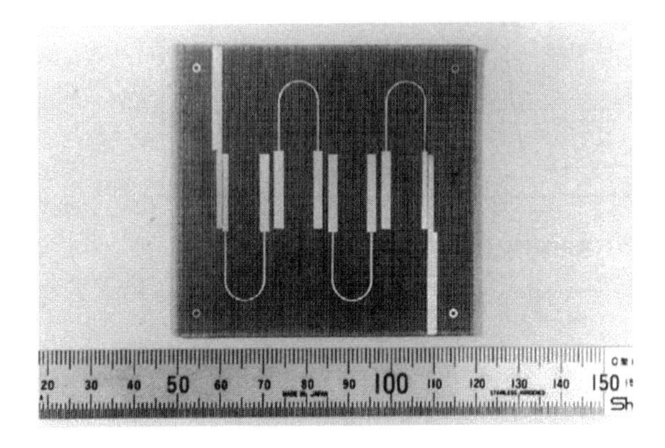

Fig. 4.7. Photograph of the experimental BPF

The line length correction $\Delta\ell$ due to fringing effects is calculated for determination of ℓ_2.

$$\Delta\ell = H \cdot \frac{\ln 2}{\pi} = 0.70\,\mathrm{mm},$$
$$\ell_2 = \ell_1 - \Delta\ell = 17.52\,\mathrm{mm}.$$

Consequently, ℓ_T is obtain as,

$$\ell_T = 2(\ell_1 + \ell_2) = 71.48\,\mathrm{mm}.$$

Line width W and line spacing S are determined by (4.20) and (4.21), these results are shown in Table 4.1. Thus all physical parameters of the experimental SIR-BPF are obtained.

Figure 4.7 shows the circuit pattern of the experimental BPF based on the above design. The unit resonators composing the filter all possess a hairpin structure [6], but spacing between branch lines of the resonator are wide

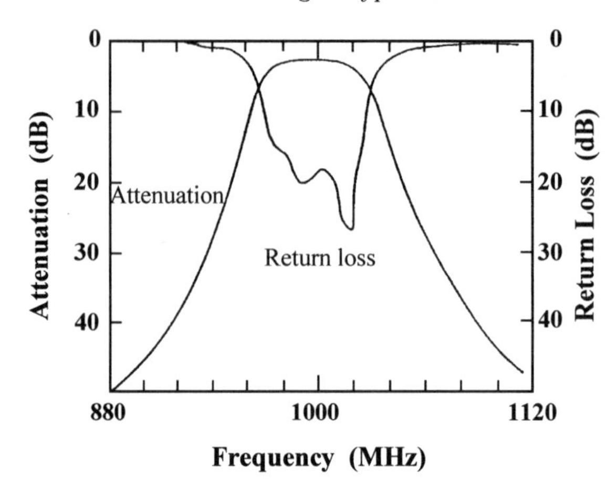

Fig. 4.8. Measured frequency responses of the experimental BPF

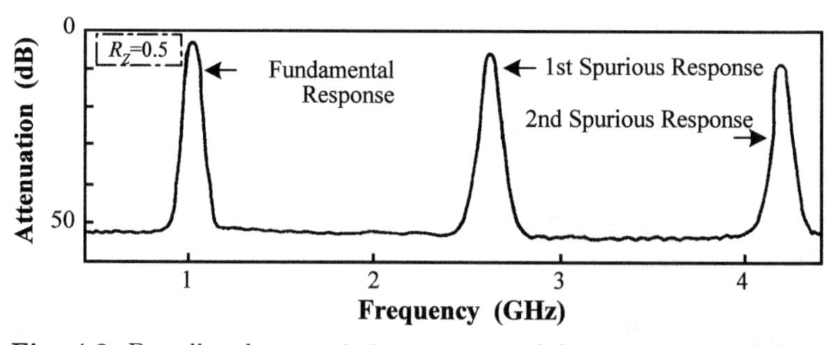

Fig. 4.9. Broadband transmission response of the experimental BPF

enough to discount unexpected effects due to coupling. Figure 4.8 shows the frequency response near pass-band of the experimental filter. These results suggest that the actual center frequency has shifted to 996 MHz and return loss characteristics have not met expected values. Such deviation can be improved by fine-tuning of each resonator, but more importantly, the obtained characteristics meet with design specifications regardless of adjustment. Insertion loss at mid-band measures 2.8 dB, and the unloaded-Q estimated from this result is about 150. For achievement of low-loss performance, an alternative design based on $R_Z > 1$ should be adopted, although doing so will sacrifice miniaturization and stop-band characteristics.

Figure 4.9 shows the broadband transmission characteristics of the experimental filter. The data illustrates the spurious responses generated at $2.6f_0$ and $4.2f_0$ which closely agree with theoretical values. This allows us to confirm that the stop-band has been expanded to $2.5f_0$. These results illustrate one of the attractive features of the SIR where spurious frequencies are generated at frequencies shifted from the integer multiples of fundamental frequency f_0. As previously described, this property proves extremely useful

Table 4.2. Comparison of the data of the trial BPF between designed and measured values

	Designed Values	Measured Values
Center Frequency	1000 MHz	996 MHz
Insertion Loss	$422/Q_0$ dB	2.8 dB
0.5 dB Bandwidth	>20 MHz	30 MHz
Attenuation (f_0 ±100 MHz)	45 dB	46 dB
Spurious Frequency	$f_{S1} = 2.55 f_0$ $f_{S2} = 4.10 f_0$	$f_{S1} = 2.60 f_0$ $f_{S2} = 4.20 f_0$

in the design of an output filter applied for suppression of harmonic components generated in nonlinear circuits. Design values and measured results of the experimental filters are summarized in Table 4.2. We conclude that good agreement has been obtained between the theoretical and measured values, thus proving the validity of the design method described in Sect. 4.1.3.

Although the above design example utilizes a dielectric substrate with low permittivity, when considering miniaturization of the BPF, the use of high permittivity materials can be extremely effective. As for low-loss characteristics, we have already stated that a design based on $R_Z > 1$ is preferable. The development of a 1.5 GHz band diplexer has been reported [7], this filter realizing a compact size and low insertion loss using a dielectric substrate of $\varepsilon_r = 89$ and SIR design of $R_Z = 1.4$.

(b) BPF Combining Different SIR Structures

Conventional resonator-coupled BPF generally apply resonators of uniform structure for design and fabrication convenience. Conversely, a combination of different SIR structures can be adopted for a special purpose such as a spurious-free BPF with wide stop-band. As an application example, here we consider a BPF with wide stop-band characteristics for spurious suppression. Such characteristics are based on the capability of controlling spurious resonance frequencies by changing the SIR structure while maintaining the same fundamental resonance frequency. As previously described, this technique is easily realized with the application of SIR, and claims that stop-band characteristics were considerably improved by applying a combination of two different structures of coaxial $\lambda_g/4$ type SIR have been reported [8].

This technique can also be applied to stripline BPF using $\lambda_g/2$ type SIR. Moreover, the suppression of spurious responses can be effectively improved by introducing distributed coupling circuits such as parallel coupled-lines,

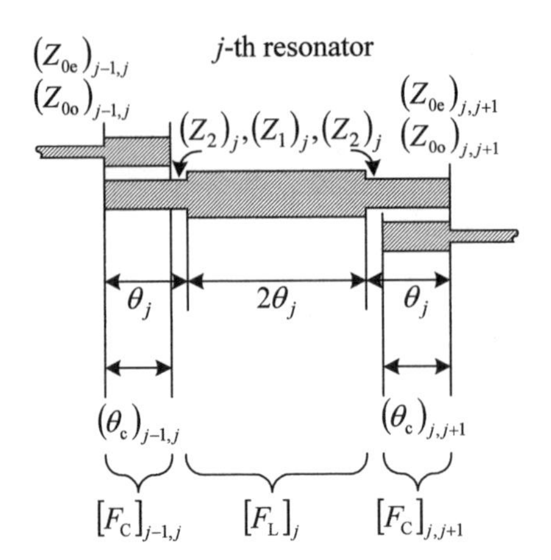

Fig. 4.10. Structure of the j-th resonator unit and its surrounding coupling sections

because the electrical performance of these circuits is intrinsically frequency dependent. In contrast to structures employing lumped-element coupling circuits, this becomes one of the attractive features of a stripline BPF applying $\lambda_g/2$ type SIR. It is well-known that antiparallel coupled-lines attain a maximum coupling strength at a coupling angle of $(1/2+n)\pi$ and a minimum value at $n\pi$, where n indicates the integer number. Making use of this property enables spurious suppression applying parallel coupled-lines [9].

We start our discussion with the calculation method of response level in the stop-band. The frequency range of the stop-band is far from the center frequency of the BPF, and thus the equivalent circuit shown in Fig. 4.4 cannot be applied to calculate response levels. This is because this circuit is valid only near the resonance frequency (center frequency) of the BPF, and thus the original circuit shown in Fig. 4.3. must be considered. Consequently, (4.6) must be considered as the F-matrix of the coupling section.

Figure 4.10 illustrates the structure of the j-th resonator unit and its surrounding coupling sections. In the figure $[F_c]_{j-1,j}$ and $[F_c]_{j,j+1}$, respectively indicate the F-matrix of the coupling section between the $(j-1)$th and j-th resonator, and j-th and $(j+1)$-th resonator, while the F-matrix for the single transmission-line section of the j-th resonator is defined as $[F_L]_j$. Expressing the SIR-BPF as a cascaded connection of these matrices, the total F-matrix of the BPF is obtained as follows:

$$\begin{aligned}[F_t] = [F_C]_{01}\,[F_L]_1\,[F_C]_{12}\,[F_L]_2 \cdots \\ \cdots [F_C]_{j-1,j}\,[F_L]_j\,[F_C]_{j,j+1} \cdots \\ \cdots [F_C]_{n-1,n}\,[F_L]_n\,[F_C]_{n,n+1}\,. \end{aligned} \tag{4.22}$$

Transmission characteristics in both the pass-band and stop-band can be obtained from the above equation.

We next describe the actual design procedure and a practical design example. The basic specifications for this experimental BPF are the same as the BPF described in (a), with an extra requirement restricting spurious responses to below -30 dB at frequencies of up to $5f_0$. Since the basic design procedures, except the stop-band requirements, are the same as the previous example, here we focus on the suppression of spurious responses in the stop-band.

Although the combination of SIR structure to be applied ranks among the first to be considered, there exists no systematized method to determine a proper combination from the specified levels of spurious response. Thus, selection of SIR structural combinations was achieved by a trial and error method, where the candidates for SIR combinations are first chosen, followed by calculations of spurious response levels based on (4.22), and finally a simple combination meeting with target specifications is selected from a practical viewpoint. Consequently, the following simple combination was adopted. The structural parameters R_Z of the four unit resonators are expressed as,

$$R_{Z1} = R_{Z4} = 1.00,$$
$$R_{Z2} = R_{Z3} = 0.50.$$

UIR, represented above by $R_Z = 1$, is applied to the first and last (fourth) resonator. Coupling angles are chosen as,

$$\theta_{01} = \theta_{56} = 90°,$$
$$\theta_{12} = \theta_{23} = \theta_{34} = 35.3°.$$

These values were tested to confirm that this configuration enables suppression of spurious response levels to below -30 dB and has few spurious response peaks. Table 4.4 shows the calculated values of spurious frequencies and spurious response levels of this BPF design. With the impedance ratio R_Z determined, we next design the resonators and coupling sections according to the previously described technique. Table 4.3 shows the design parameters of the coupled-lines.

Table 4.3. Design parameters of coupled-lines

$j, j+1$	$(\theta_c)_{j,j+1}$ (deg)	$J_{j,j+1}/Y_0$	$(Z_{0e})_{j,j+1}$ (Ω)	$(Z_{0o})_{j,j+1}$ (Ω)
01	90.0	0.2969	69.3	39.6
12	35.3	0.0601	55.8	45.3
23	35.3	0.0391	53.6	46.8
34	35.3	0.0601	55.8	45.3
45	90.0	0.2969	69.3	39.6

Table 4.4. Calculated and measured values of spurious responses

Spurious Freq.	$2f_0$	$2.55f_0$	$3f_0$	$4f_0$	$4.1f_0$	$5f_0$	$5.1f_0$
Calculated Level (dB)	< -100	-34.5	-43.5	-60	-45.7	-79.5	< -100
Measured Level (dB)	< -55	-33	-46	-50	< -55	< -55	< -55

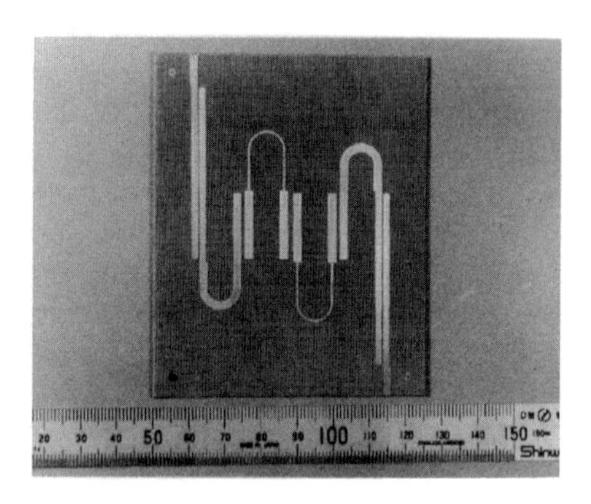

Fig. 4.11. Photograph of the experimental BPF for spurious suppression

Figure 4.11 shows the circuit pattern of the experimental BPF designed on the same substrate as in the previous example. Calculated and measured transmission characteristics near the center frequency of the pass-band are shown in Fig. 4.12. The measured data is in close agreement with design, and although return-loss characteristics show some decline, this can be improved by fine-tuning of each resonator. Figure 4.13 illustrates stop-band characteristics, where spurious responses are observed at $2.6f_0$, $3.0f_0$, and $4f_0$, and the corresponding response levels are -33 dB, below -45 dB, and below -45 dB, respectively. These characteristics sufficiently meet with specifications, enabling the stop-band to be expanded to $5f_0$, and thus demonstrating the validity of the design method described here.

Table 4.4 shows a comparison between the theoretical values and measured values of the spurious frequencies and response levels of the BPF. The first and fourth resonator being UIR ($R_Z = 1.0$), spurious resonances are generated at the integer multiples of the center frequency $2f_0, 3f_0, 4f_0, 5f_0, \cdots$. The second and third resonators ($R_Z = 0.50$) possess spurious response at $2.55f_0, 4.10f_0, 5.10f_0 \cdot \cdots$. Although there are seven spurious resonance frequencies below $5f_0$, only three distinct spurious responses are observed at $2.55f_0, 3f_0, 4f_0$. These results also match the simulated values.

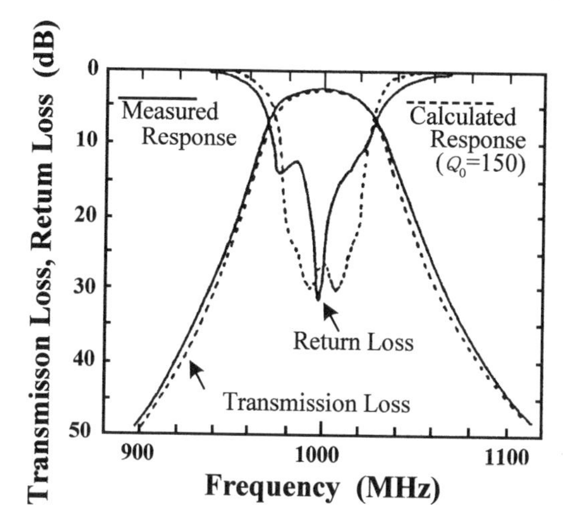

ig. 4.12. Calculated and meaured responses of the experiental BPF

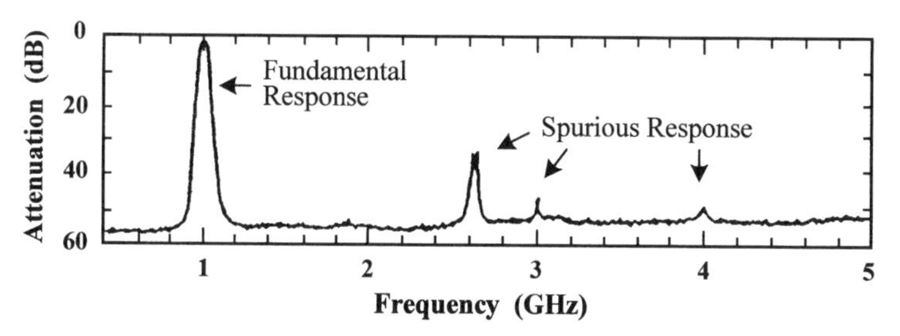

Fig. 4.13. Broadband transmission response of the experimental BPF for spurious suppression

The distinct spurious response at $2.55f_0$ can be explained as the synergistic effect of the second and third resonator being resonant at that point, and the coupling strength between the two taking maximum strength due to a coupling angle of $90°(= 35.3 \times 2.55)$ at $2.55f_0$. On the contrary, the first and fourth resonators are resonant at $2f_0$, where there exist no coupling between the resonators because the coupling angle becomes $180°(= 90 \times 2)$, and thus the response level is suppressed. These results imply that it is effective to utilize the frequency dependence of the coupling section, as well as the combination of resonator structures, in order to obtain maximum spurious response suppression and stop-band expansion of the BPF, and for this purpose the stripline SIR is a highly reliable resource.

4.2 Internally Coupled SIR

4.2.1 Basic Structures and Resonance Condition

Although stripline resonators are the most frequently used resonator structure in microwave circuit design, they possess a fundamental disadvantage of large size. In an attempt to overcome this problem, the authors have proposed a new resonator structure, composed of a ring-shaped stripline possessing two open-ends connected by a capacitor as shown in Fig. 4.14, which has since been developed and applied to RF and microwave devices [10,11]. This resonator structure, called the stripline split-ring resonator, enables miniaturization while maintaining an attractive feature of the $\lambda_g/2$ resonator, namely a structure that needs no DC-grounded points in the stripline. It is extremely important in the design of practical RF circuits that the applied resonators have no grounded points. From an electrical performance viewpoint the existence of a DC-grounded point is apt to ruin circuit stability by inducing stray inductance and increasing circuit losses. Moreover, from a manufacturing view point, such grounded points require an additional process for forming via-holes, which will degrade reliability and cost-efficiency of the device.

An internally coupled SIR structure replaces the lumped-element capacitor C_T shown in Fig. 4.14 with a distributed coupling circuit [12]. We start our discussion on this topic with the resonance conditions of the circuit shown in Fig. 4.14, which is derived from the input admittance Y_i at the end of the transmission line as shown in the figure. The circuit in Fig. 4.14 can be equivalently expressed as a four-terminal circuit shown in Fig. 4.15. Thus, the split-ring resonator can also be regarded as a four-terminal circuit consisting of a parallel-connected single transmission line and capacitor, with an open-circuited output terminal. The F-matrix of the serial connected C_T and single transmission line, respectively expressed as $[F_1], [F_2]$, are defined as follows.

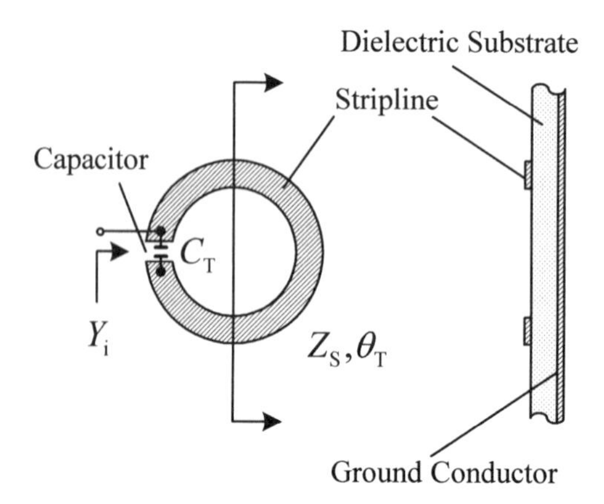

Fig. 4.14. Basic structure of a stripline split-ring resonator

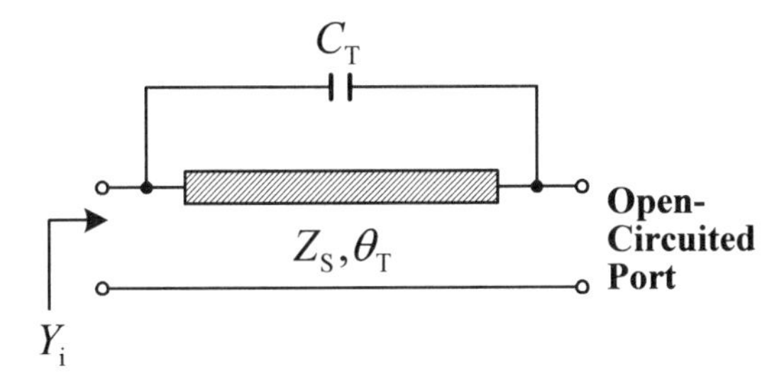

Fig. 4.15. An equivalent circuit of a split-ring resonator

$$[F_1] = \begin{bmatrix} A_1 & B_1 \\ C_1 & D_1 \end{bmatrix} = \begin{bmatrix} 1 & 1/\mathrm{j}wC_\mathrm{T} \\ 0 & 1 \end{bmatrix}, \tag{4.23}$$

$$[F_2] = \begin{bmatrix} A_2 & B_2 \\ C_2 & D_2 \end{bmatrix} = \begin{bmatrix} \cos\theta_\mathrm{T} & \mathrm{j}Z_\mathrm{S}\sin\theta_\mathrm{T} \\ \mathrm{j}\sin\theta_\mathrm{T}/Z_\mathrm{S} & \cos\theta_\mathrm{T} \end{bmatrix}. \tag{4.24}$$

$[F_\mathrm{T}]$, defined as the total F-matrix of a parallel connection between the above circuits, is obtained as,

$$[F_\mathrm{T}] = \begin{bmatrix} A_\mathrm{T} & B_\mathrm{T} \\ C_\mathrm{T} & D_\mathrm{T} \end{bmatrix}, \tag{4.25}$$

where

$$A_\mathrm{T} = \frac{A_1B_2 + A_2B_1}{B_1 + B_2},$$

$$B_\mathrm{T} = \frac{B_1B_2}{B_1 + B_2},$$

$$C_\mathrm{T} = \frac{-(A_1 - A_2)(D_1 - D_2) - (B_1 + B_2)(C_1 + C_2)}{B_1 + B_2},$$

$$D_\mathrm{T} = \frac{B_1D_2 + B_2D_1}{B_1 + B_2} = A_\mathrm{T}.$$

Taking the load impedance of the circuit shown in Fig. 4.15 as Z_L, input admittance Y_i is obtained from (4.25) as,

$$Y_\mathrm{i} = \frac{C_\mathrm{T}Z_\mathrm{L} + D_\mathrm{T}}{A_\mathrm{T}Z_\mathrm{L} + B_\mathrm{T}} = \frac{C_\mathrm{T} + D_\mathrm{T}/Z_\mathrm{L}}{A_\mathrm{T} + B_\mathrm{T}/Z_\mathrm{L}} \tag{4.26}$$

In the case of a split-ring resonator, Z_L is considered infinite, thus enabling Y_i to be obtained from (4.26) as,

$$Y_\mathrm{i} = \frac{C_\mathrm{T}}{A_\mathrm{T}} = \frac{-(A_1 - A_2)(D_1 - D_2) - (B_1 + B_2)(C_1 + C_2)}{A_1B_2 + A_2B_1} \tag{4.27}$$

Substituting each matrix element from (4.23) and (4.24), we obtain,

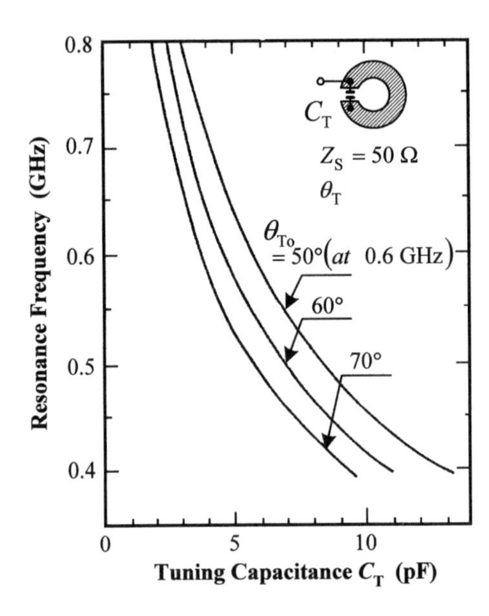

Fig. 4.16. Calculated results of resonance frequency of a split-ring resonator

$$
\begin{aligned}
Y_{\mathrm{i}} &= \frac{-(1-\cos\theta_{\mathrm{T}})^2 + \mathrm{j}\sin\theta_{\mathrm{T}}\cdot(\mathrm{j}Z_{\mathrm{S}}\sin\theta_{\mathrm{T}} + 1/\mathrm{j}\omega C_{\mathrm{T}})/Z_{\mathrm{S}}}{\mathrm{j}Z_{\mathrm{S}}\sin\theta_{\mathrm{T}} + \cos\theta_{\mathrm{T}}/\mathrm{j}\omega C_{\mathrm{T}}} \\
&= \mathrm{j}Y_{\mathrm{S}}\cdot\frac{Y_{\mathrm{S}}\sin\theta_{\mathrm{T}} - 2\omega C_{\mathrm{T}}(1-\cos\theta_{\mathrm{T}})}{Y_{\mathrm{S}}\cos\theta_{\mathrm{T}} - \omega C_{\mathrm{T}}\sin\theta_{\mathrm{T}}}
\end{aligned}
\tag{4.28}
$$

The resonance condition of a shunt-resonator can be expressed as $Y_{\mathrm{i}} = 0$, and thus the following equation can be derived.

$$
Y_{\mathrm{S}}\sin\theta_{\mathrm{T}} - 2wC_{\mathrm{T}}(1-\cos\theta_{\mathrm{T}}) = 0. \tag{4.29}
$$

Figure 4.16 illustrates the relationship between C_{T} and resonance frequency when $Z_{\mathrm{S}} = 50\,\Omega$ and line length is constant, where results suggest that resonance frequency can be controlled by simply changing capacitance C_{T}. The application of this feature to wide-band electrically tunable BPF have been reported [10].

The input impedance Z_{i} of the split-ring resonator can also be obtained from the denominator of (4.28), and from this the serial resonance condition is expressed as follows:

$$
Y_{\mathrm{S}}\cos\theta_{\mathrm{T}} - wC_{\mathrm{T}}\sin\theta_{\mathrm{T}} = 0. \tag{4.30}
$$

The resonance points obtained from (4.29) and (4.30) are in the relation of resonance and antiresonance, and are known to exist at a small distance, while their relative frequency span can be further reduced by applying a greater C_{T} value. This implies that a split-ring resonator can be designed to inductively operate in a narrow frequency band, where reactance shows an extremely rapid change. Such properties are ideal for the reactance element of oscillators, and hence low phase noise oscillators employing split-ring resonators

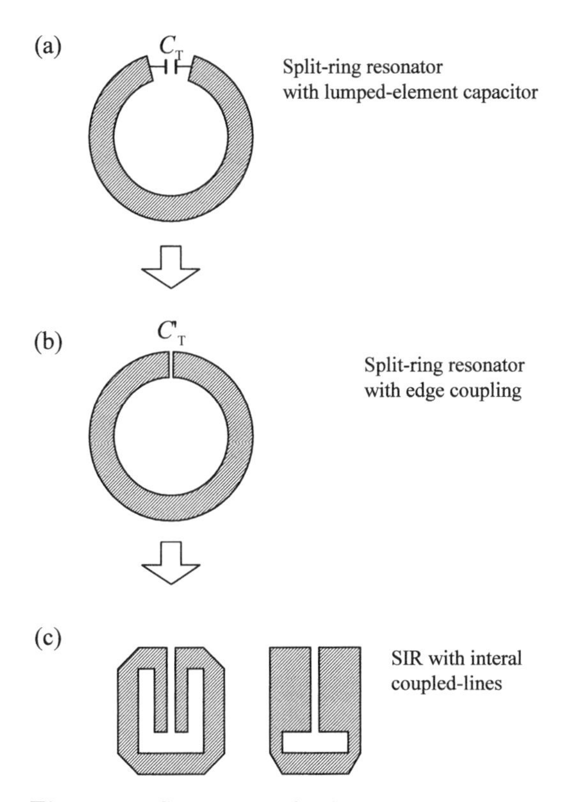

Fig. 4.17. Structures of split-ring resonator and related SIR. (**a**) Conventional split-ring resonator. (**b**) Edge-coupled split-ring resonator. (**c**) SIRs with internal coupling

have been put into practical use in the UHF band for mobile communication equipment.

Split-ring resonators employing lumped-element capacitors have advantages of small size and the capability of controlling resonance frequency, while possessing simultaneous disadvantages of larger circuit losses and larger resonance frequency variance due to the lumped-element capacitor C_{T}. Such drawbacks can be conquered by replacing the lumped-element capacitor C_{T} with distributed coupling elements, as shown in Fig. 4.17. Figure 4.17b indicating a structure based on edge coupling is one such example, but capacitance values realized by this configuration are too small to design a compact split-ring resonator. Thus, two structures in Fig. 4.17c are desirable for practical applications. Resonators of such a structure are regarded as SIR possessing internal coupling elements realized by parallel coupling of the transmission lines near both open-ends. Applying this structure enables a compact SIR design while simultaneously increasing design flexibility.

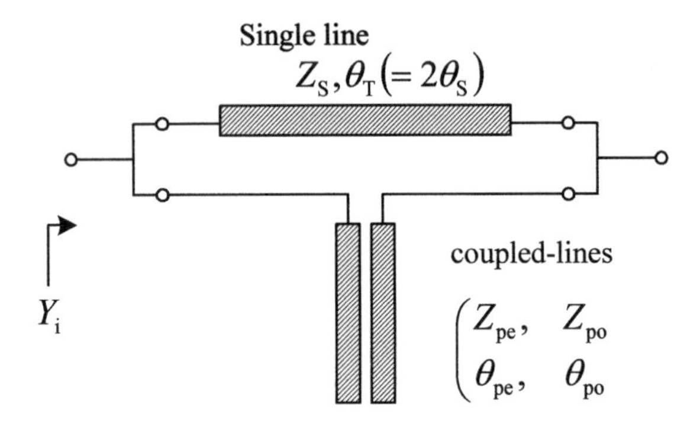

Fig. 4.18. An equivalent circuit of SIR with internal coupled-lines

Next, let us discuss the resonance frequency of the resonator structures shown in (c). Fig. 4.18 illustrates an equivalent circuit expression of this structure, and as the circuit in Fig 4.15, this circuit can be analyzed as a parallel connection of a single transmission-line and parallel coupled-lines with open-ends [12]. Even- and odd-mode impedance and corresponding coupling angle of parallel coupled-lines are represented by Z_{0e}, Z_{0o} and θ_{pe}, θ_{po}, respectively. The F-matrix corresponding to (4.23) can be expressed as,

$$[F_1] = \begin{bmatrix} A_1 & B_1 \\ C_1 & D_1 \end{bmatrix} = \begin{bmatrix} \dfrac{Z_{pe}\cot\theta_{pe} + Z_{po}\cot\theta_{po}}{Z_{pe}\cot\theta_{pe} - Z_{po}\cot\theta_{po}} & -j\dfrac{2Z_{pe}Z_{po}\cot\theta_{pe}\cot\theta_{po}}{Z_{po}\cot\theta_{pe} - Z_{po}\cot\theta_{po}} \\ j\dfrac{2}{Z_{pe}\cot\theta_{pe} - Z_{po}\cot\theta_{po}} & \dfrac{Z_{pe}\cot\theta_{pe} + Z_{po}\cot\theta_{po}}{Z_{pe}\cot\theta_{pe} - Z_{po}\cot\theta_{po}} \end{bmatrix}. \tag{4.31}$$

Resonance conditions are obtained by substituting the matrix elements of (4.24) and (4.31) to (4.27), and solving Y_i. Consequently, the resonance condition results as follows.

$$\begin{aligned} &\left(Z_{pe} \cdot Z_{po}\cot\theta_{pe}\cot\theta_{po} - Z_S^2\right)\sin\theta_T \\ &+Z_S\left(Z_{pe}\cot\theta_{pe} + Z_{po}\cot\theta_{po}\right)\cos\theta_T \\ &-Z_S\left(Z_{pe}\cot\theta_{pe} - Z_{po}\cot\theta_{po}\right) = 0. \end{aligned} \tag{4.32}$$

The phase velocities of even- and odd-modes are again assumed equal for simplicity as $\theta_{pe} = \theta_{po} = \theta_p$, and thus (4.32) is expressed as,

$$\begin{aligned} &\left(Z_{pe} \cdot Z_{po}\cot\theta_p - Z_S^2\tan\theta_p\right)\sin 2\theta_S \\ &+Z_S\left(Z_{pe} + Z_{po}\right)\cos 2\theta_S - Z_S\left(Z_{pe} - Z_{po}\right) = 0, \end{aligned} \tag{4.33}$$

where $\theta_S = \theta_T/2$. Furthermore, parameters Z_P and C, defined as follows, are introduced.

$Z_P \equiv \sqrt{Z_{pe} \cdot Z_{po}}$: geometric means of even- and odd-mode impedance

$C \equiv (Z_{pe} - Z_{po}) / (Z_{pe} + Z_{po})$: coupling coefficient

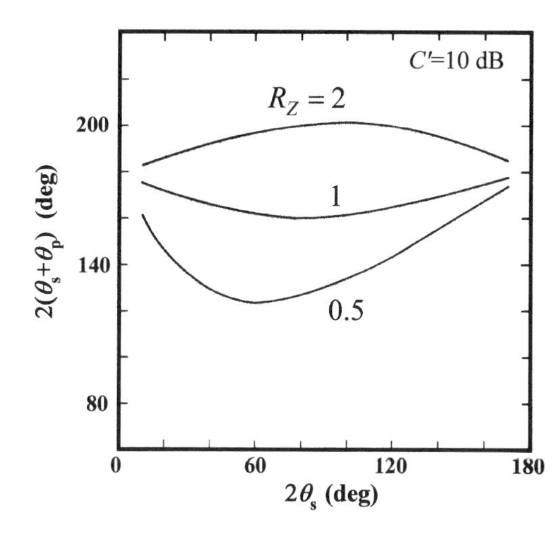

Fig. 4.19. Calculated results of resonance condition when taking R_Z as parameter

Using these parameters, we obtain

$$Z_{\mathrm{pe}} + Z_{\mathrm{po}} = 2Z_{\mathrm{P}}/\sqrt{1-C^2},$$
$$Z_{\mathrm{pe}} - Z_{\mathrm{po}} = 2CZ_{\mathrm{P}}/\sqrt{1-C^2}.$$

Substituting the above equations equation into (4.33), the final resonance condition is obtained as,

$$\sqrt{1-C^2} \cdot (R_{Z\mathrm{p}} \cot\theta_{\mathrm{p}} - \tan\theta_{\mathrm{p}}/R_{Z\mathrm{p}}) \cdot \sin 2\theta_{\mathrm{S}} + 2\cos 2\theta_{\mathrm{S}} - 2C = 0, \tag{4.34}$$

where $R_{Z\mathrm{p}} = Z_{\mathrm{P}}/Z_{\mathrm{S}}$: impedance ratio.

Based on these results, the relationship between total resonator length $2\,(\theta_{\mathrm{S}} + \theta_{\mathrm{P}})$ and single line length is obtained to examine the conditions of miniaturization. Figures 4.19 and 4.20 show resonance conditions when taking R_Z and C' as parameters, respectively. Here C' indicates the coupling coefficient in dB ($C' = 20\log(1/C)$). Results indicate that a lower R_Z and a greater C (or smaller C') are required for miniaturization. In addition, the conditions yielding minimum resonator length $2(\theta_{\mathrm{S}} + \theta_{\mathrm{p}})$ for a constant R_Z and C are given as $\theta_{\mathrm{S}} = \theta_{\mathrm{p}}$, which is equivalent to that of a conventional SIR without coupled-lines.

4.2.2 Equivalent Circuits at Resonance

The resonance conditions expressed as (4.32) and (4.34) are general descriptions derived from the input admittance of the resonator. However, their relationship is too complicated to allow understanding of their physical meanings. Thus, in this section we focus on an equivalent circuit expression at resonance to investigate such resonance conditions [12]. A $\lambda_{\mathrm{g}}/2$ type resonator with open-ends possesses a maximum electric field distribution near

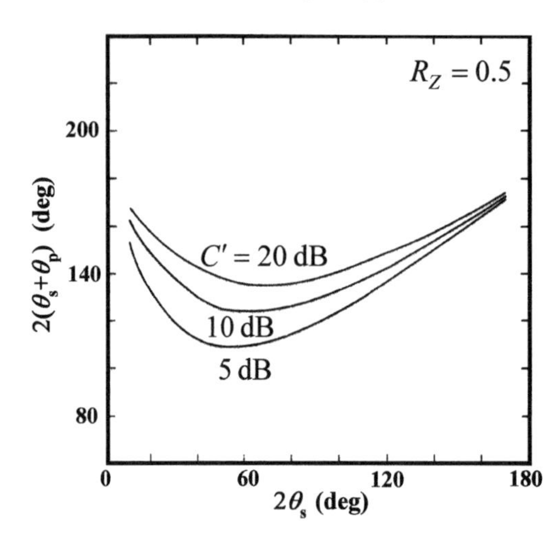

Fig. 4.20. Calculated results of resonance condition when taking C' as parameter

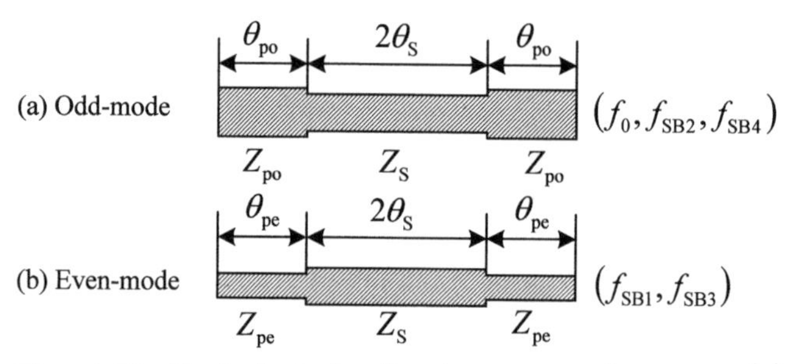

Fig. 4.21. Equivalent circuits at resonance frequency. (**a**) Odd-mode case. (**b**) Even-mode case

both open-circuited ends at resonance, and the phase difference between both ends is π(radian) at fundamental resonance f_0 and 0(radian) at the following higher mode of resonance $2f_0$. This implies that a SIR with internal coupled-lines generates an odd-mode field distribution at fundamental resonance f_0, and thus only Z_{po} need be taken into consideration as the impedance of the parallel coupled-lines. Consequently, the equivalent circuit at resonance can be expressed as a conventional single line SIR as shown in Fig. 4.21a. Similarly, the equivalent circuit for the following higher resonance mode $(2f_0)$ is expressed as Fig. 4.21b because the electric field distribution in this case will be even-mode. Thus, an SIR with coupled-lines can be expressed as a conventional SIR composed of two single transmission lines, hence resonance conditions are given as,

$$\text{Even-mode}: \tan\theta_S \cdot \tan\theta_{po} = R_{Zo} = Z_{po}/Z_S, \tag{4.35}$$

$$\text{Odd-mode}: \tan\theta_S \cdot \tan\theta_{pe} = R_{Ze} = Z_{pe}/Z_S. \tag{4.36}$$

The equivalent circuits described above are enough to facilitate analysis near resonance frequencies, and furthermore provide a good understanding of circuit behavior at resonance. For off-resonance frequencies, however, circuit performance can only be examined by applying (4.32).

4.2.3 Filter Design Examples

(a) Ring-Shaped SIR-BPF

As a MIC applicable filter structure, we discuss a ring-shaped SIR-BPF [12] illustrated in Fig. 4.22. A micro-stripline configuration is adopted for the transmission-line, while basic properties of the substrate are given as,

Relative dielectric constant	: $\varepsilon_r = 10.5$
Loss tangent	: $\tan\delta_d = 0.002$
Substrate thickness	: $H = 1.27\,\text{mm}$.

To obtain a generalized design method, design charts are prepared in accordance with the techniques described in the appendix. The basic structure of the resonator is given as,

$$\begin{aligned} Z_S &= 50\,\Omega, \\ 2\theta_S &= 100^\circ \qquad \text{(at 1.5 GHz)}, \\ Z_{pe} &= 97.6\,\Omega, \\ Z_{po} &= 29.5\,\Omega. \end{aligned}$$

Figure 4.23 shows the relationship between electrical length θ_p of the coupling section and resonance frequency. The figure suggests that fine-tuning of frequency can be achieved by adjusting the length of the parallel coupled-lines in the manufacturing process.

Unloaded-Q(Q_0) and external-Q(Q_e) are next to be obtained. Deriving an analytical expression for Q_0 and Q_e becomes extremely difficult due to the existence of internal distributed coupling within the resonator, and thus the method described in the appendix is introduced. Here a simple coupling structure by tapping is adopted as an external coupling circuit. Figures 4.24

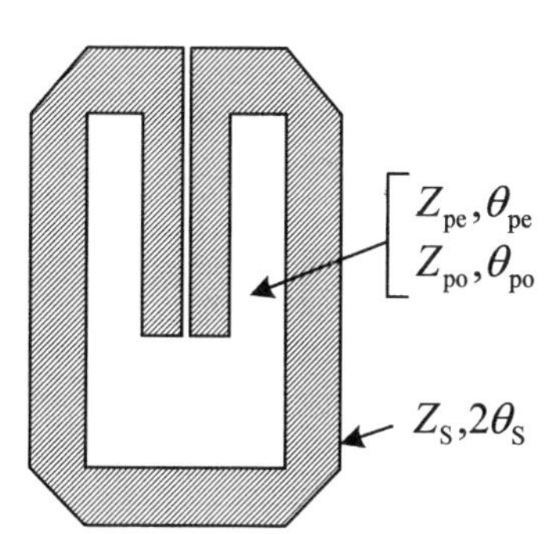

Fig. 4.22. Electrical parameters of a ring-type SIR with internal coupled-lines

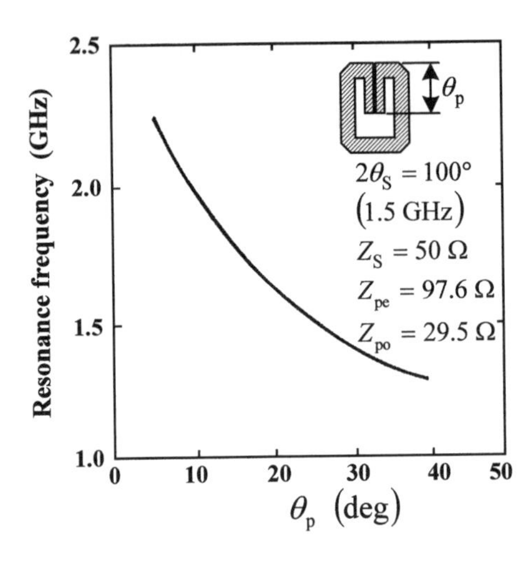

Fig. 4.23. Filter design chart: Resonance frequency

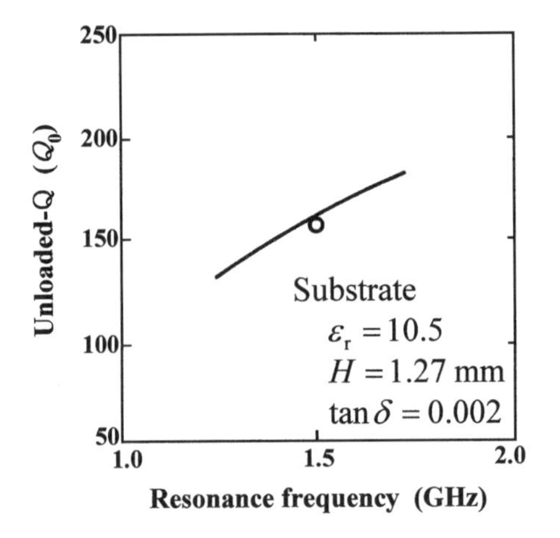

Fig. 4.24. Filter design chart: Unloaded-Q (Q_0)

and 4.25 illustrate the results for Q_0 and Q_e, respectively. The interstage coupling circuits between resonators are realized by parallel coupled-lines, and the coupling coefficient k is obtained by the method described in the appendix. Figure 4.26 shows calculated results of k.

Thus, by applying these design charts shown in Figs. 4.23 to 4.25, it is possible to design a BPF with specified characteristics in the 1.5 GHz band. As a design example, we consider a BPF specified as follows: center frequency $f_0 = 1.5\,\text{GHz}$, pass-band bandwidth $W > 30\,\text{MHz}$, and attenuation $L_S > 20\,\text{dB}$ (at $f_0 \pm 0.1\,\text{GHz}$). Filter design parameters meeting the above specifications are determined as follows:

Number of stages	:	$n = 3$
Input and output coupling	:	$Q_e = 19.68$
Interstage coupling	:	$k_{12} = k_{23} = 0.0433.$

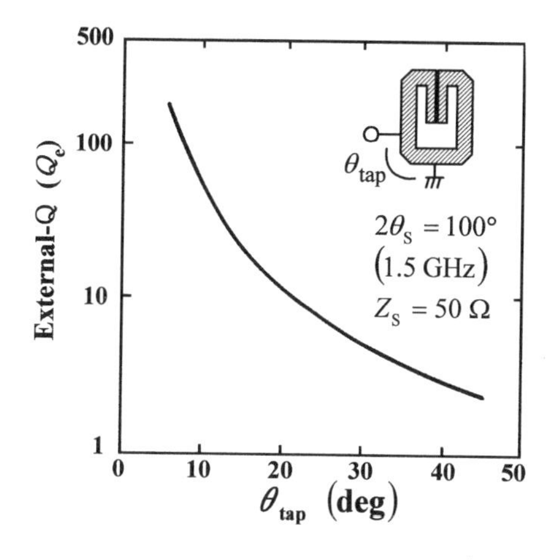

Fig. 4.25. Filter design chart: External-Q (Q_e)

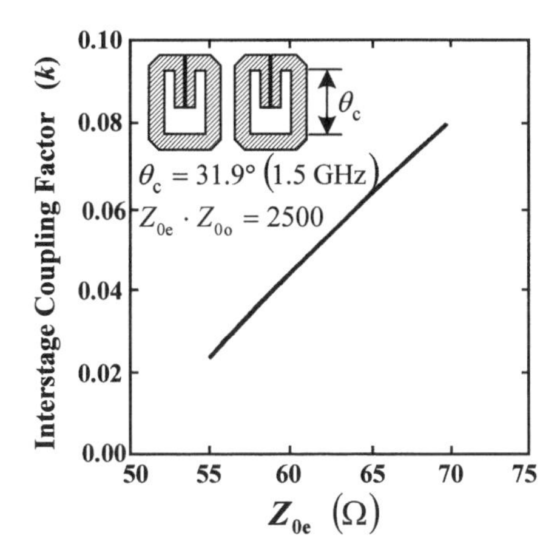

Fig. 4.26. Filter design chart: Inter-stage coupling factor

Consequently, the electrical parameters of the BPF are obtained from the design charts as:

resonator parameters:

$$Z_S = 50\,\Omega, \quad \theta_S = 100°,$$
$$Z_{pe} = 97.6\,\Omega, \quad Z_{po} = 29.5\,\Omega, \quad \theta_p = 26°,$$

coupling parameters:

$$\text{input and output} \quad \theta_{tap} = 13.8°,$$
$$\text{interstage} \quad Z_{0o} = 43.7\,\Omega, Z_{0e} = 57.2\,\Omega, \theta_c = 31.9°.$$

A photograph of the experimental BPF is illustrated in Fig. 4.27, while measured characteristics are illustrated in Fig. 4.28. Figure 4.27 compares the ring-shaped SIR-BPF (below) to a conventional hairpin resonator filter

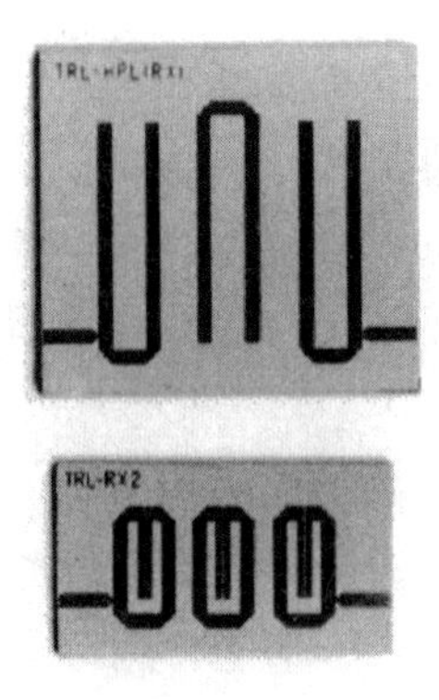

Fig. 4.27. Photograph of the experimental BPF. *Above*: Conventional hairpin resonator filter. *Below*: Ring-type SIR filter

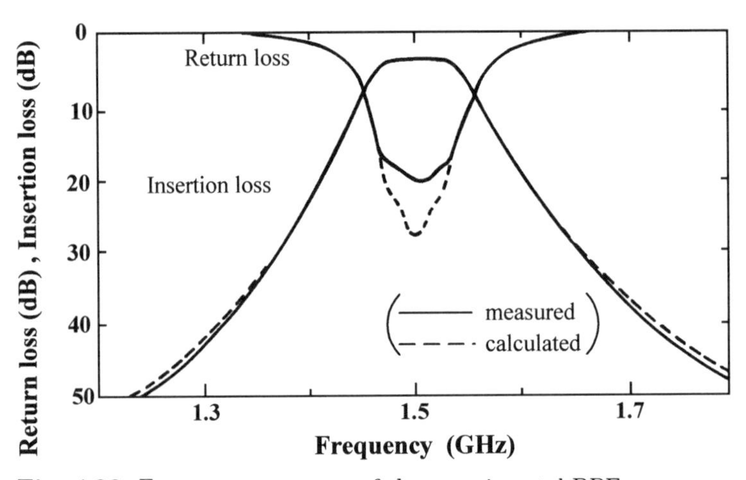

Fig. 4.28. Frequency responses of the experimental BPF

(above), where we see a size reduction of more than 50% by applying the ring-shaped structure. The two filters are based on the same design specifications, and measured electrical performance also proved to be equivalent. In addition, the measured data in Fig. 4.28 agree well with the designed data, thus proving the validity of the described design method.

(b) Hairpin-shaped SIR-BPF

Figure 4.29 illustrates the basic structure and electrical parameters of the hairpin-shaped SIR to be discussed. Compared with ring-shaped SIR, this

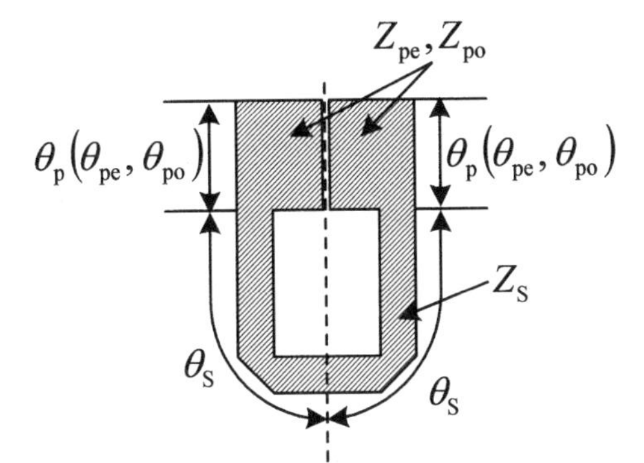

Fig. 4.29. Electrical parameters of hairpin-shaped SIR

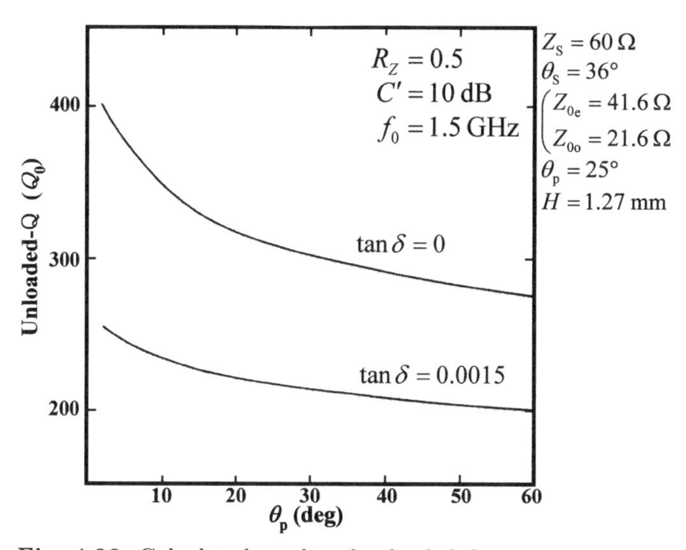

Fig. 4.30. Calculated results of unloaded-Q

resonator structure features an advantage of low impedance coupling between the coupled-lines.

As with structure in (a), we prepare design charts for a 1.5 GHz band BPF by applying the method described in the appendix. The applied substrate has a dielectric constant of 10.5, thickness of 1.27 mm, and a loss tangent of 0.015, which is slightly smaller than the substrate used in (a). Figure 4.30 illustrates calculated results of one such design example. The results suggest that unloaded-Q primarily depends on the loss tangent $\tan\delta_d$ of the substrate, while the coupling angle θ_p of the coupled-lines also contributes to change. The effects of the coupling angle θ_p are presumably due to the loss generated in the distributed coupling section, but more likely brought about by unloaded-Q degradation caused by the shortening of total resonator length.

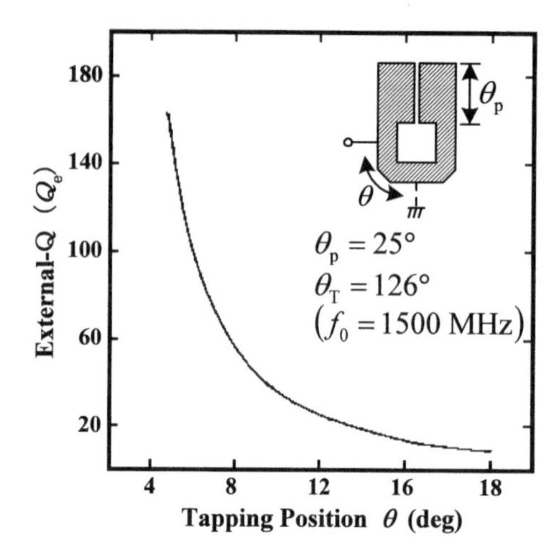

Fig. 4.31. Filter design chart: External-Q

This can be understood from the figure, where the unloaded-Q degrades in accordance with a gain in θ_p even in the case of $\tan\delta_d$=0. Electric field loss caused by loss tangent $\tan\delta_d$ being the dominant factor for losses generated at parallel coupled-lines, the unloaded-Q should not be dependent on θ_p if $\tan\delta_d$=0. Thus, we understand that the influence of size effects, namely the total resonator length, is dominant.

Figure 4.31 shows the relationship between tapping position and external $Q(Q_e)$. The electrical parameters of the resonator are given as,

$$Z_S = 60\ \Omega, \quad \theta_S = 36°,$$
$$Z_{0e} = 41.6\,\Omega, \quad Z_{0o} = 21.6\,\Omega, \quad Z_{0p} = 25°.$$

These values imply the structure's suitability for application to narrow band filters with a relative bandwidth of several to several tens of per cent. Figure 4.32 shows calculated examples of interstage coupling factor k as a function of resonator spacing G, in the case of a parallel-coupling configuration. These results reveal that this configuration is suitable for a narrow band filter with a bandwidth of less than several per cent. On the contrary, when a broadband filter with strong coupling is required, a structure possessing antiparallel coupling between the resonators is considered desirable. Based on the above discussions, we design a BPF with the following specifications which are equivalent to that of (a).

Center frequency	: $f_0 = 1500\,$MHz
Passband bandwidth	: $W > 30\,$MHz
Attenuation	: $L_S > 20\,$dB (at $f_0 \pm 100\,$MHz).

As previously discussed, this filter can be designed by adopting the filter parameters as,

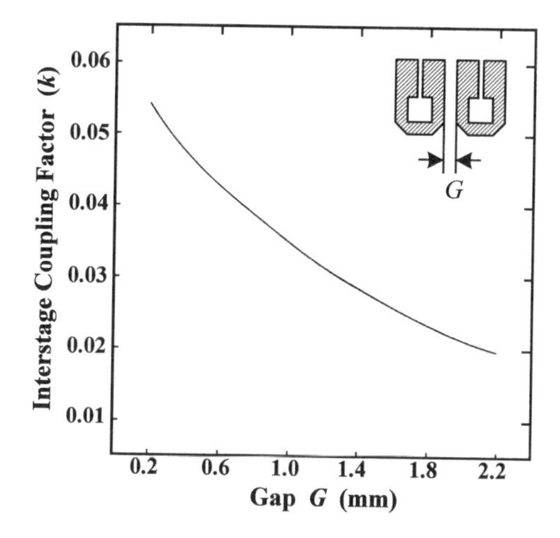

Fig. 4.32. Filter design chart: Interstage coupling factor

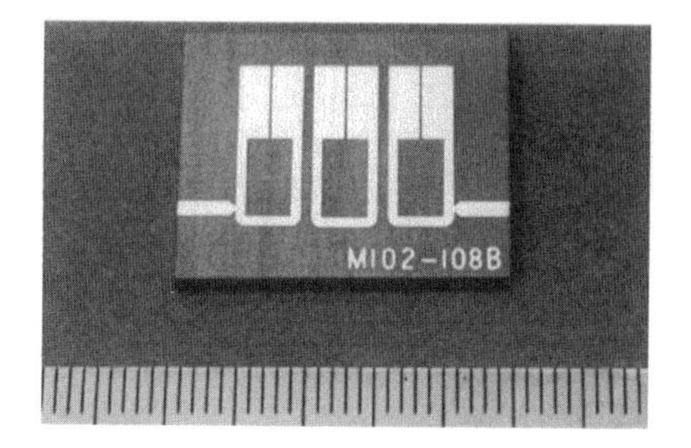

Fig. 4.33. Photograph of the experimental filter

Number of stages : $n = 3$
Coupling parameters : $Q_\mathrm{e} = 19.7, k_{12} = k_{23} = 0.0434$.

The resonator parameters are given previously, and thus the coupling parameters are obtained from the design charts of Figs. 4.31 and 4.32 as,

Tapping position : $\theta_\mathrm{p} = 13.6°$
Resonator spacing : $G = 0.58\,\mathrm{mm}$.

From the above discussion, all physical parameters of the experimental BPF are determined. Figure 4.33 shows a photograph of the experimental filter. The applied substrate size measures 23 mm×17 mm, which is nearly equivalent to the ring-shaped SIR-BPF described in (a). The measured characteristics are shown in Fig. 4.34. These results have been obtained without frequency adjustments of the resonators. The insertion loss at mid-band measured 2.5 dB, corresponding to the designed value estimated for an assumed unloaded-Q of 210. Furthermore, excluding a slight disturbance in the higher frequency bands, measured frequency responses also indicate close agreement with design.

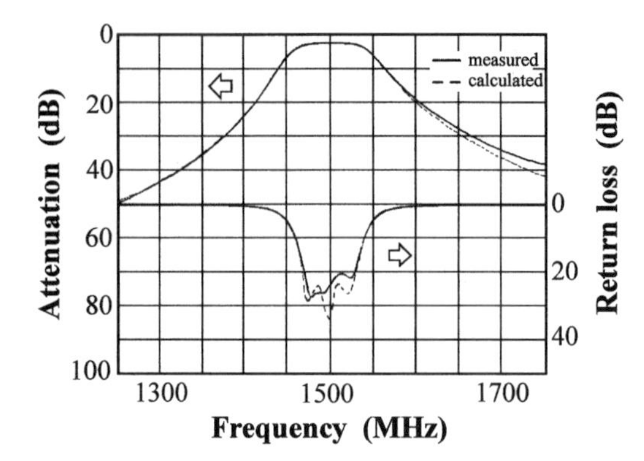

Fig. 4.34. Transmission characteristics of the experimental filter

(c) Application to High-T_c Superconductor Filters

Ever since their discovery in 1986, attempts to realize practical application of high critical-temperature superconducting (HTS) thin film materials have been tested in various industrial fields, and much attention is paid to its practical application to microwave filters for communication systems [13]. One promising application in this field is to the receiver front-end filter for base stations in mobile communication systems, where a low insertion loss and steep attenuation are required for reception sensitivity and selectivity. The HTS filter possesses a potential possibility of realizing a dramatic size reduction for communication filters which require narrow band and extremely low insertion loss properties. This is based on expectations of extremely small surface resistivity of HTS thin film, resulting in an ultrahigh unloaded-Q value even for a reduced resonator size. The drawback is the necessity of a cryogenic environment, where device miniaturization becomes the key factor for reduction of operating power and improvement of system reliability. Thus, the miniaturization of the HTS filter itself becomes a key issue for practical application [14–16].

Table 4.5 summarizes measured surface resistivity at 10 GHz for typical HTS materials [17], namely $YBa_2Cu_3O_x$ and $Tl_2Ba_2CaCu_2O_x$. Results show HTS surface resistivity values reduced to 1.4 and 3.3 per cent of metal copper

Table 4.5. Characteristics of superconducting materials measured at 10 GHz

Material	T_c(K)	Surface Resistivity (mΩ)		
		300K	77K	50K
$YB_2Cu_3O_x$	94	—	0.3	0.08
$Tl_2Ba_2Cu_2O_x$	110	—	0.13	0.09
Cu	—	26	9.0	4.2

values. In addition, in the lower frequency bands the surface resistivity of HTS material is further reduced in comparison to conventional metals, due to a proportional relationship to frequency squared. Typical dielectric substrates applied to HTS thin film include single crystal dielectric material such as MgO and $LaAlO_3$, which possess a loss tangent of approximately 7×10^{-6} at 10 GHz, and reports claim unloaded-Q value of over 10^5 for HTS resonator applying these substrate materials in the microwave region.

As previously described, the HTS filter is suitable for a filter with narrow-band characteristics and a steep attenuation. As an example, an experimental filter based on Chebyschev response with center frequency $f_0 = 2.0\,\text{GHz}$, relative bandwidth $w = 0.01$, number of stages $n = 8$, and pass-band ripple $R = 0.01\,\text{dB}$ is discussed. The pass-band insertion loss is given by (3.37) as,

$$L_0 \approx \frac{4434}{wQ_0} \sum_{j=1}^{n} g_j \text{dB} = \frac{5640}{Q_0} \text{dB}.$$

The unloaded-Q value of a conventional stripline resonator is dependent on structure, but maximum values rarely exceed several hundred at 2 GHz. This implies an insertion loss L_0 of several dozen dB, which will not be tolerated in a practical application. However, by applying HST materials to this structure, an unloaded-Q of over 10^4 can easily be achieved, thus allowing for a practical insertion loss value of less than 0.5 dB.

Figure 4.35 shows two filter circuit patterns designed by the above specifications. The substrate material is $LaAlO_3(\varepsilon_r = 24)$ with a thickness H of 0.5 mm, while the HTS material is $YBa_2Cu_3O_x$. In the figure, (a) and (b) indicates a BPF using hairpin-shaped SIR and a BPF using conventional parallel-coupled UIR, respectively. In the case of (a), the geometric mean of the even- and odd-mode impedance of the parallel coupled-lines is set at 25 Ω and the characteristic impedance of the single line is 50 Ω. Adopting a

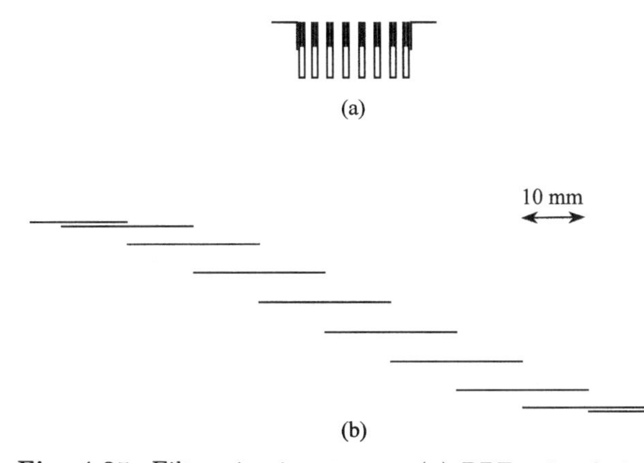

Fig. 4.35. Filter circuit patterns. (**a**) BPF using hairpin-shaped SIRs. (**b**) BPF using conventional parallel-coupled UIRs

hairpin-shaped SIR structure enables the total size of the experimental filter to be reduced to $15 \times 25\,\mathrm{mm}^2$. The filter size in the case of (b) is $40 \times 105\,\mathrm{mm}^2$, thus implying a size reduction of 90% by effectively utilzing the advantages of the hairpin-shaped SIR [16]. Figure 4.36 shows a photograph of the experimental filter mounted on a test jig, while performance is shown in Fig. 4.37. Measured insertion loss in the pass-band was less than 0.4 dB. This includes losses generated in the connectors and measurement cables, which are difficult to separate in practical terms. Though it is hard to accurately determine the unloaded-Q value, an unloaded-Q value exceeding 10^5 can be assumed when considering the results obtained from a single resonator. Practical application of HTS filters will require technologies reaching out to various fields such as materials, circuit design, and thermodynamics, and various problems remain still to be solved. However, the HTS filter possesses the possibility of offering totally new devices, and high hopes are placed on future progress to practical application.

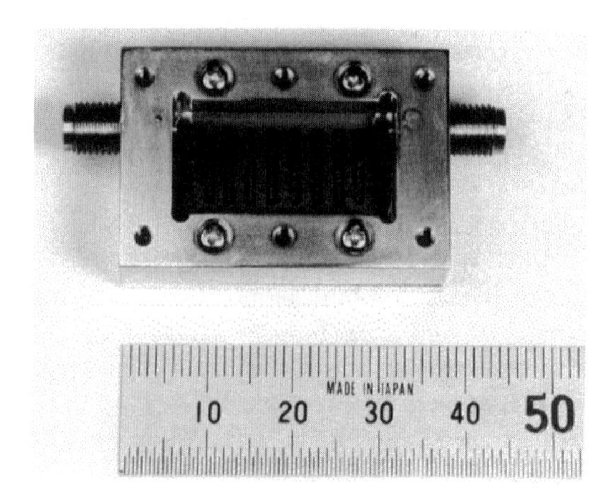

Fig. 4.36. Photograph of the experimental high T_c superconductor BPF, by courtesy of Dr. Enokihara

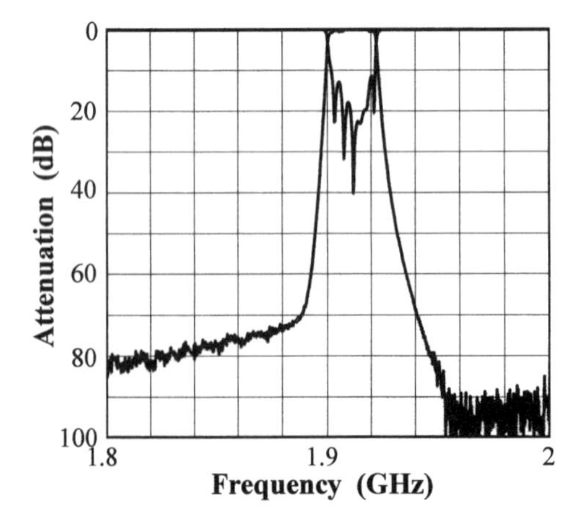

Fig. 4.37. Measured frequency responses of the experimental BPF

4.2.4 Application to Oscillator and Mixer Circuits

(a) Input Impedance Characteristics

As described in Sect. 4.2.1, the split-ring resonator possesses resonance and antiresonance points which are alternately aligned on the frequency axis. This phenomenon is also seen in a $\lambda_g/2$ type SIR possessing internal couplings. Figure 4.38 illustrates the calculated impedance of a hairpin-shaped SIR. The shunt resonance, namely antiresonance, frequency of this resonator is designed at 1.5 GHz, and the electrical parameters are given as $Z_S = 60\,\Omega$, $\theta_S = 38°$, $Z_{pe} = 41.6\,\Omega$, $Z_{po} = 21.6\,\Omega$, $\theta_p = 25°$.

As can be seen from the figure, the resonance (serial resonance) points and antiresonance (shunt resonance) points, shown respectively as $\mathrm{Im}[Z_i] = 0$ and $\mathrm{Im}[Z_i] = \infty$, are alternately located. The lowest antiresonance point is indicated as f_{P1} in the figure, and this resonance point is utilized for BPF design. In the case of an oscillator application, a resonator is usually applied as an inductive element, and thus the frequency range between antiresonance f_{P1} and resonance point f_{S1}, where the impedance of the resonator becomes inductive, is employed. This span becomes relatively narrow for a strong internal coupling or a small impedance ratio R_Z of the hairpin-shaped SIR, thus resulting in a steep change in frequency response of the reactance component, which consequently contributes to the reduction of oscillator phase noise.

(b) Potential Distribution in the SIR

Figure 4.39 shows the potential distribution along the transmission line of the resonator when assuming the voltage of the open-circuited end of the hairpin-shaped SIR to be V_0. The dotted line indicated by ① corresponds

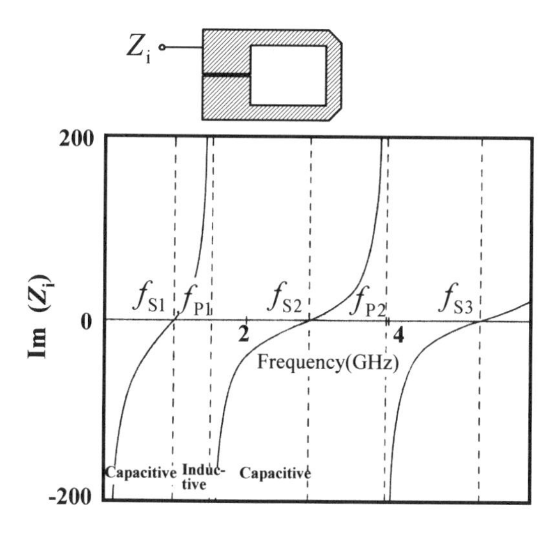

Fig. 4.38. Calculated input impedance of a hairpin-shaped SIR

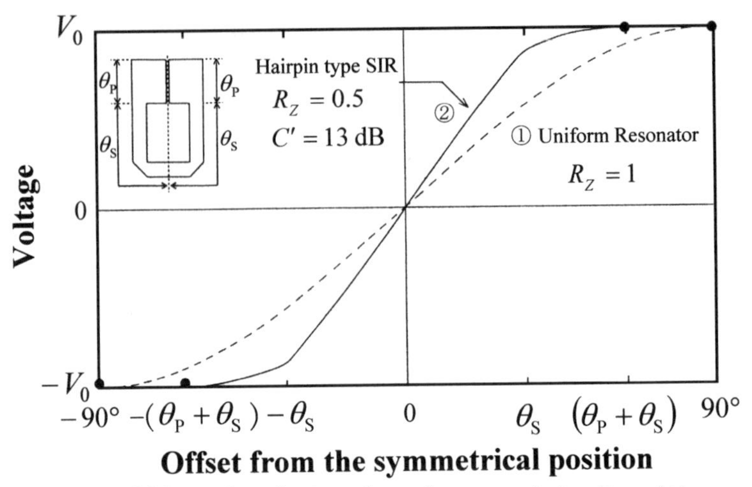

Fig. 4.39. Voltage distribution along the transmission line of the resonator

to a conventional half-wavelength resonator, showing a sinusoidal potential distribution. The solid line indicated by ② is the case of a hairpin-shaped SIR. This distribution is not sinusoidal, and has a tendency of slight change in the coupled-lines section and steep change in the single line section. However, the voltages at the two open-ends still become $+V_0$ and $-V_0$, behaving in reverse phase to each other. These characteristics are a basic property required for a push-push oscillator described later. Furthermore, the center of the single line section becomes a virtual RF short-circuited point where potential attains a minimum value and current attains a maximum value. For this section of the SIR it is preferable to provide magnetic coupling with an output circuit.

(c) Circuit Configuration of Oscillators and Mixers

Figure 4.40 shows the basic configuration of an oscillator using a hairpin-shaped SIR. This circuit consists of a SIR and an active circuit which generate negative resistance by employing an active device such as a transistor, connected to one of the open-circuited ends of the SIR. This active circuit

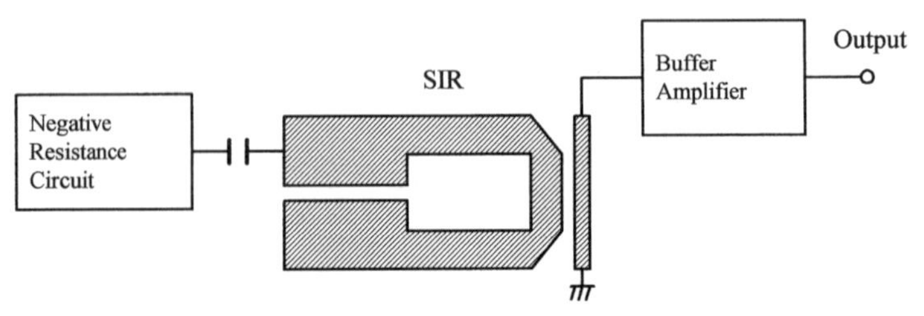

Fig. 4.40. Basic circuit configuration of an oscillator using hairpin-shaped SIR

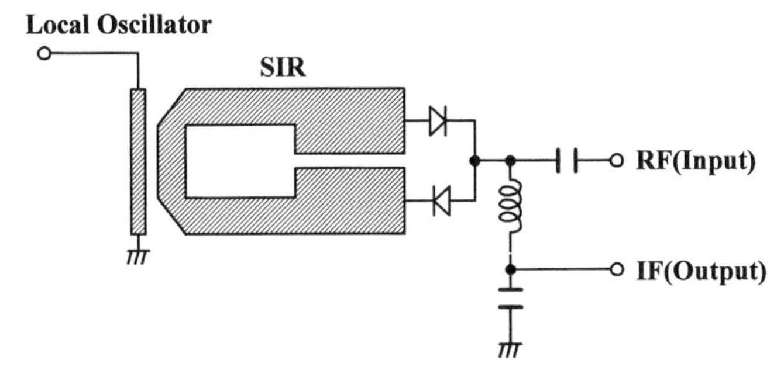

Fig. 4.41. Basic circuit configuration of a single balanced mixer

possesses a capacitive component as well as negative resistance, and thus the oscillating frequency is determined by the condition at which the inductive component of the resonator compensates the capacitive component of the active circuit. As previously described, the frequency range in which a SIR operates inductively is extremely narrow, and thus the reactance component shows a rapid change within this range, consequently contributing to the frequency stability of an oscillator applying hairpin-shaped SIR. Power can be extracted from the circuit by magnetic coupling at the center of the SIR where the current obtains a maximum value. In this example, the output power is passed on through the buffer amplifier for circuit stability. Fine-tuning of the oscillator can be achieved by trimming the conductive film at the open-ends of the SIR or adding a minute lumped-element serial capacitor to the coupled-lines.

Figure 4.41 illustrates the basic configuration of a balanced mixer using a hairpin-shaped SIR [18]. This mixer circuit can effectively drive the diode pair by utilizing the high voltages of opposite phase generated at the two open-ends of the SIR. The SIR is tuned to the input frequency of the local oscillator as shown in the figure, and local oscillator power is injected to the SIR by magnetic coupling at the symmetrical plane of the SIR to obtain circuit balance. From a circuit theory point of view, the SIR in this configuration can be considered as a balance-to-unbalance transformer (balun) inserted in the local oscillator port. The RF input is connected to the center point of the serial connected diode pair as shown in the figure to enable wide-band operation which is usually required for RF signals. The IF output is obtained from the same point through a LPF composed of LC circuits. Advantages of this mixer include its planar configuration of the single-balanced mixer and miniaturization of the balun due to the application of hairpin-shaped SIR, and its wide applicable frequency range from RF to microwave.

(d) Experimental Push-Push Oscillator and Its Characteristics

Practical applications of SIR as oscillator resonant elements have been described in the previous section. Such applications are based on SIR features of narrow inductive frequency span, which consequently realize oscillators with low phase noise characteristics. Here we consider an application to a push-push oscillator [20] that utilizes the inverse voltage characteristics at the two open-ends of the SIR. The push-push oscillator consists of two identical negative resistance circuits that are separately connected to the two open-ends of the SIR, realizing two oscillating circuits that operate at uniform frequency while possessing a reverse phase. Figure 4.42 illustrates the basic circuit structure based on a hairpin-shaped SIR. This oscillating circuit is composed of two negative circuits and one SIR, and this configuration enables single operation of each oscillator as well as simultaneous operation of the two oscillators. The negative resistance circuits are composed of identical components, and thus both oscillating frequencies are nearly equal. Therefore, for simultaneous operation, the injection-lock phenomenon of an oscillator enables the oscillating frequency to be locked onto a unique frequency determined by the more stable signal among the two circuits. In this injection-locked state, the output signals of the two oscillating circuits encounter reverse phase.

Due to nonlinear operations, the output signal of a typical oscillator contains various high level harmonics other than the fundamental component. Although the output signals of the two oscillators show reverse phase for the fundamental component, for the second harmonics they operate in common phase. This makes it possible to combine the power of the second harmonics by applying an in-phase hybrid, while simultaneously suppressing the fundamental component. Consequently, this configuration can be applied to realize a high efficiency osciplier, which is an oscillator with a multiplier function, that does not need an output filter. By contrast, by applying a 180° hybrid the fundamental signals of the two oscillators can be combined while suppressing the second harmonics. Oscillator phase noise reduction near resonance can effectively be realized with this configuration, for, while the power

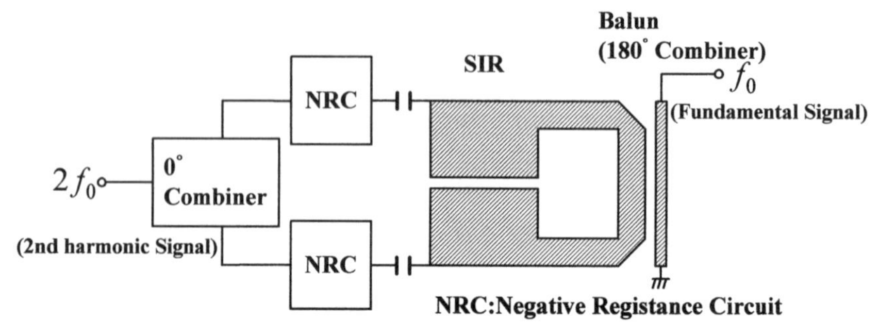

Fig. 4.42. Basic circuit configuration of the experimental push-push oscillator

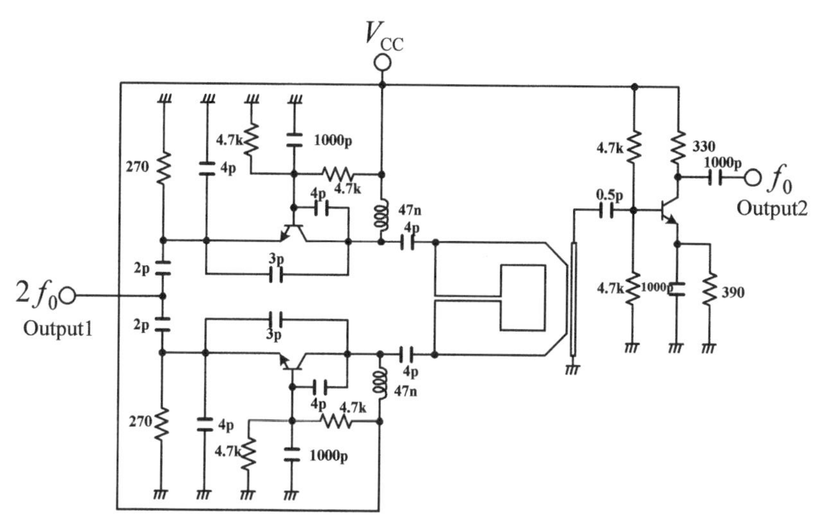

Fig. 4.43. Circuit diagram of the experimental push-push oscillator

Fig. 4.44. Photograph of the experimental push-push oscillator

of both oscillators are amplified, the lower frequency noise components which are mixed into both oscillators in common phase are cancelled by the 180° hybrid.

Figure 4.43 shows the circuit diagram of the experimental 750 MHz band push-push oscipllator. A Colpitz type oscillating circuit has been adopted, and the electrical parameters of the hairpin-shaped SIR are as follows,

$$Z_S = 63\,\Omega,\ \theta_S = 45°,$$
$$Z_{pe} = 39\,\Omega,\ Z_{po} = 25\,\Omega,\ \theta_P = 20°.$$

The parallel coupled-lines of the fundamental signal output must be designed at an adequate length in order to increase output power, and thus the single line length θ_S is increased. Figure 4.44 shows a photograph of the experimental push-push oscillator. Characteristics of the applied dielectric substrate are: $\varepsilon_r = 4.7$, substrate thickness $H = 0.8\,\text{mm}$, and $\tan\delta_d = 0.02$. This allows for a SIR unloaded-Q of about 90. Figure 4.45 shows the spectrum

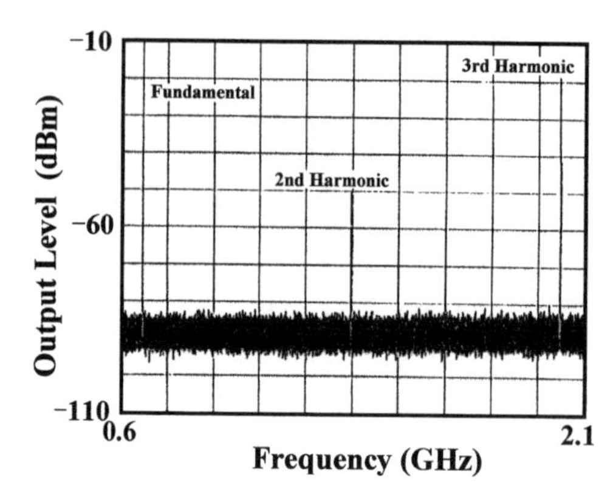

Fig. 4.45. Frequency spectrum of the experimental push-push oscillator at port-2

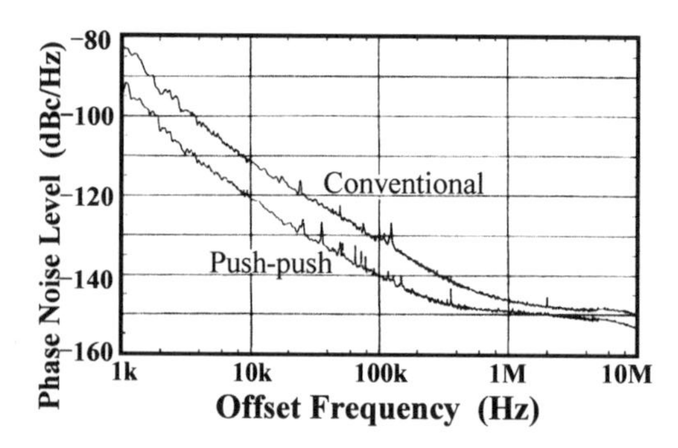

Fig. 4.46. Phase noise characteristics of the experimental oscillator

obtained from the combined output of the fundamental signals recorded at output port-2. An approximate 40 dB suppression is observed for the second harmonics as compared to the fundamental signal. Additionally, for an opposite configuration applying an in-phase combiner, results showed a 25 dB suppression of the fundamental signal as compared to the extracted second harmonics.

Figure 4.46 illustrates the phase noise characteristics of the combined output of the fundamental signal. The figure shows a comparison between a conventional Colpitz type oscillator and a push-push oscillator using SIR of the same structure as described above. Results suggest a phase noise improvement of about 9 dB by adopting the push-push oscillator configuration. These results illustrate an important feature of the push-push oscillator.

Although these experiments were carried out in the 750 MHz band to assess the fundamental characteristics of the push-push oscillator, further circuit balance can be realized by adopting planar circuits as the resonator structure while introducing MMICs for the oscillator section. These conditions are easily achieved at higher frequencies, thus implying that a push-push oscillator applying a hairpin-shaped SIR realizes maximum availability in the microwave region rather than the RF band.

5. One-Wavelength-Type SIR

A one-wavelength resonator formed by a stripline, micro-stripline, or slot-line structure features low radiation-loss characteristics, and a structural advantage enabling the elimination of parasitic components which are usually induced at the open- and short-circuited ends of conventional half- or quarter-wavelength resonators. These features account for their frequent application to standard resonators for measuring stripline characteristics and dielectric substrate properties. However, when considering their practical application as single mode resonators, the overall size of the λ_g(one-wavelength) resonator becomes a large handicap as compared to the other candidates.

It is well-known, however, that there are two orthogonal resonance modes within a one-wavelength ring resonator [1]. Using these two modes, their application to filtering devices has been well studied and practical use has also progressed [2–4]. Practical application of this dual-mode ring resonator can broadly be divided into two approaches. The first approach is an application to four-port devices by independently operating the two orthogonal modes, while the second approach is to utilize an internal coupling between the two degenerate modes to realize two-port devices [5].

We start our discussions by investigating the two orthogonal resonance modes in the ring resonator, illustrating the possibilities of exciting two modes of different frequency while maintaining spatial orthogonal conditions. In other words, the basic properties of two orthogonal resonators with different resonance frequencies are discussed. Resonator structures using SIR follow, and discussions focus on possible application to diplexers as four-port devices. Filtering devices are the primary two-port device application. We consider the analysis method of coupling between orthogonal modes, and furthermore describe design examples of filters using λ_g-type SIR.

5.1 Orthogonal Resonance Modes in the Ring Resonator

The dominant resonant mode excited in a one-wavelength type ring resonator is treated as a quasi-TEM mode for a structure with a narrow line width compared to ring radius, or a TM_{110} mode otherwise. Here we concentrate on the basic behavior of the ring resonator by introducing the simplified model based on a quasi-TEM mode resonance.

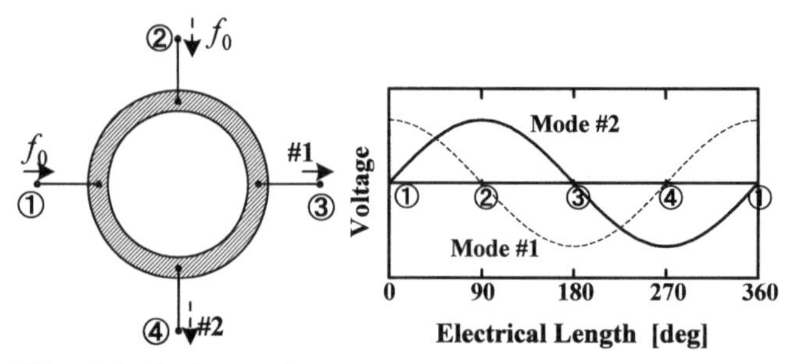

Fig. 5.1. Orthogonal resonance modes in the ring resonator

As shown in Fig. 5.1, four ports are connected to the ring resonator, each spatially separated at 90° intervals. When port① is excited with an electric field, a resonant mode illustrated as mode# 1 in the figure is generated within the resonator. The electrical potentials of port② and port④ are zero in this case, thus making it possible to simultaneously excite port② to obtain a mode#2 resonance. The electrical potentials of port① and port③ become zero for this mode, thus enabling two independent resonance modes to simultaneously exist in the ring resonator. The above-described phenomena can qualitatively be explained by the concept of the travelling wave. The incident wave at port① generates clockwise and counterclockwise travelling waves. These two waves reach port③ in the same phase, while at port② and port④ they encounter opposite phase. Therefore, the incident wave at port① can propagate to port③ while disappearing at port② and port④. In the same way, the incident wave at port② can propagate to port④ but will not appear at port① and port③. Note that this discussion is based on orthogonal modes of the same frequency. Orthogonal modes of two different frequencies will be investigated in the following chapter.

A high voltage with a 180° phase shift is generated between port① and port③ in Fig. 5.1, and by serially connecting capacitor C between these two ports, the resonance frequency can be lowered to f_ℓ ($f_0 > f_\ell$) as illustrated in Fig. 5.2a. Figure 5.2b shows an equivalent circuit expression of this circuit at resonance [5], where shunt capacitors $2C$ are connected to both ports. Circuit symmetry is preserved even after inserting capacitor C, and thus when exciting port②, the electrical potential of port① and port③ will still be zero, allowing the resonance frequency to remain at f_0. This allows for the ring resonator to possess two orthogonal resonance modes of different frequency. Utilizing such orthogonal resonance modes within a ring resonator, numerous microwave devices have been proposed and developed.

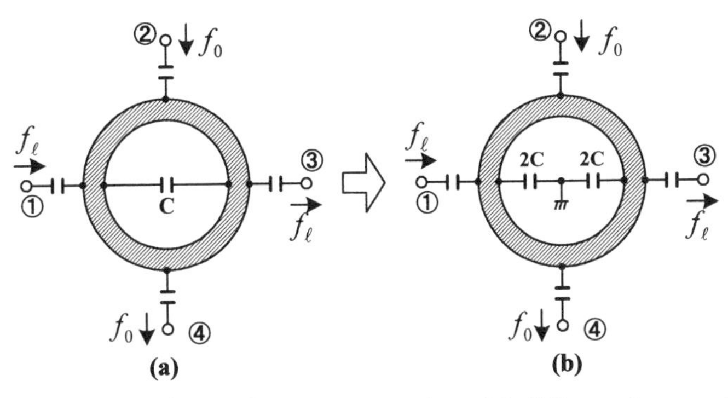

Fig. 5.2. Orthogonal resonance modes with different frequencies

5.2 Application of λ_g-Type SIR as Four-Port Devices

Figure 5.3 shows a typical application examples of a four-port device which independently utilizes orthogonal resonance modes of the same frequency. In these circuits, port① and port④ are used as input and output terminals, while external circuits are inserted between port② and port③. These structures function as filtering or oscillating devices by independently utilizing the two resonant modes. By applying a phase shifter consisting of a capacitor and/or inductor as an external circuit, a two-stage bandpass filter can be realized by optimizing circuit parameters.

A tuned amplifier has been experimentally constructed [5] by introducing a unilateral amplifier as the external circuit. This circuit consists of a pre-filter, amplifier with adequate input and output matching, and a post-filter. Conventional active filters frequently discussed in the RF and microwave region show practical performance in gain, but possess drawbacks concerning noise figure and circuit stability. Although this ring resonator applied tuned amplifier structure is not categorized as an active filter, its functions equal such active filters, thus applying a practical solution considering noise figure and circuit stability.

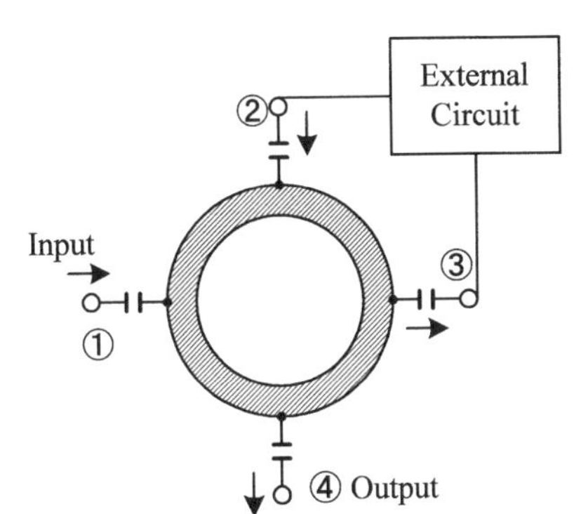

Fig. 5.3. Ring resonator four-port device with an external circuit

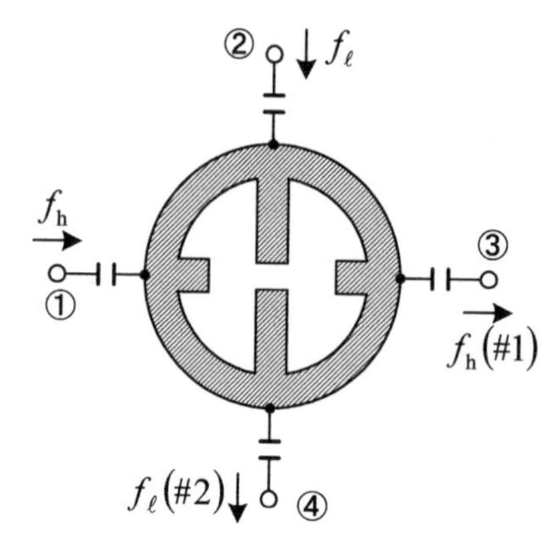

Fig. 5.4. Miniaturized dual-mode ring resonator with different frequencies

The application of a four-port ring resonator as an oscillator resonant element has been proposed. One such example is a low phase noise voltage-controlled oscillator, where an actual design has achieved electrical performance applicable to actual mobile communication equipment. Another example is the application to an osciplier, i.e. an oscillator possessing multiplier functions. Circuit behavior of this circuit can be explained using the ring resonator shown in Fig. 5.1. By realizing an oscillator structure where an active circuit possessing negative resistance is connected to port③, the potential at ports ② and ④ become nearly zero for fundamental frequency f_0. For the second harmonics $2f_0$, however, the potential attains maximum values with equal phase. Thus, by combining the output power at ports ② and ④ with capacitive or electrical coupling, a $2f_0$ output signal with suppressed fundamental components can be obtained. Reports claim an extremely stable output and a fundamental suppression level of over 18 dB for an experimental voltage controlled oscillator with a fundamental frequency of 800 MHz [5].

The principle of circuit behavior enabling two orthogonal modes within a ring resonator to resonate at different frequencies has been described in Fig. 5.2. For practical application, a planar circuit configuration possessing open stubs at each port as shown in Fig. 5.4 is preferable when considering factors such as circuit performance degradation due to lumped-elements and requirements of size reduction. Furthermore, a four-port ring resonator with stepped impedance configuration, as shown in Fig. 5.5 can be applied in order to satisfy requirements for a wider separation of the resonance frequencies and to avoid restriction of stub use. Expressing the resonance frequencies exciting ports ① and ②, respectively, as f_{01} and f_{02}, the equivalent circuits at resonance for both orthogonal modes can be expressed as in Figs. 5.6a and b. Since both circuits are composed of two parallel connected $\lambda_g/2$ type SIR, the discussion in Sect. 4.1.1 can be applied and thus the resonance condition can be expressed as follows [6]:

$$\tan\theta_1 \tan\theta_2 = Z_2/Z_1 = R_Z .$$

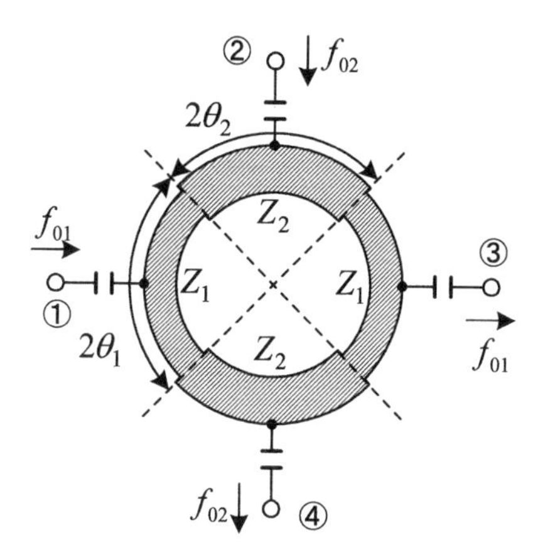

Fig. 5.5. Stepped impedance dual-mode ring resonator with different resonance frequencies

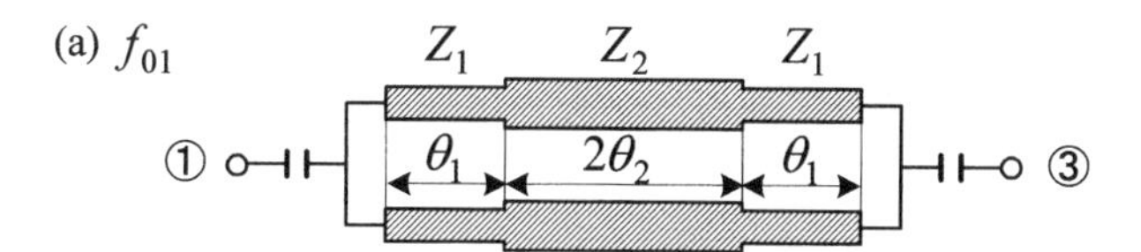

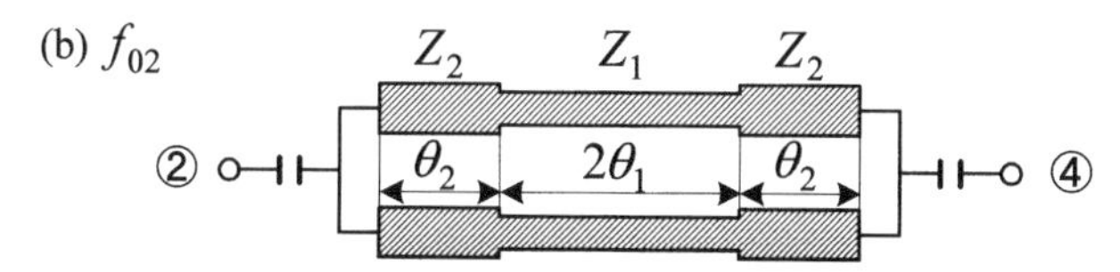

Fig. 5.6. Equivalent circuits of two resonance modes of ring-type SIR. (**a**) $R_Z = Z_2/Z_1 > 1$. (**b**) $R_Z = Z_2/Z_1 < 1$

Under the conditions of $\theta_1 = \theta_2 = \theta$, assuming the θ values at f_{01} and f_{02} as θ_{01} and θ_{02}, respectively, we obtain,

$$\theta_{01} = \tan^{-1}\sqrt{R_Z}$$

$$\theta_{02} = \tan^{1}\sqrt{1/R_Z}.$$

Furthermore, assuming $\theta_0(= \pi/4)$ as the θ value of the uniform impedance resonator (UIR),

$$\theta_{01} = \frac{f_{01}}{f_0}\theta_0 = \frac{\pi}{4}\frac{f_{01}}{f_0}$$

$$\theta_{02} = \frac{f_{02}}{f_0}\theta_0 = \frac{\pi}{4}\frac{f_{02}}{f_0}.$$

Thus, by normalizing the frequency difference Δf between f_{01} and f_{02} by f_0 we obtain,

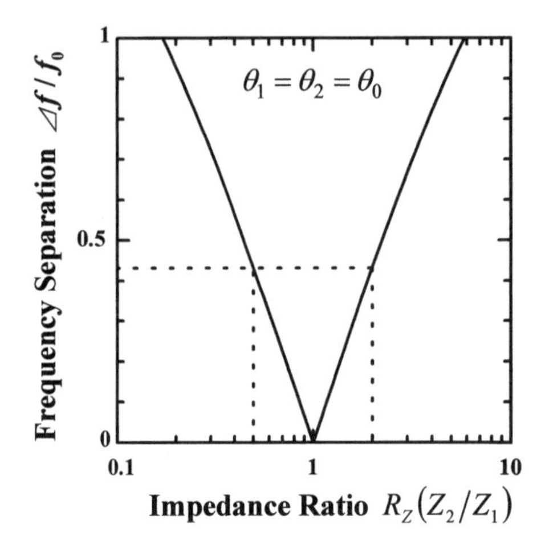

Fig. 5.7. Resonance frequency separation of a ring type dual-mode SIR

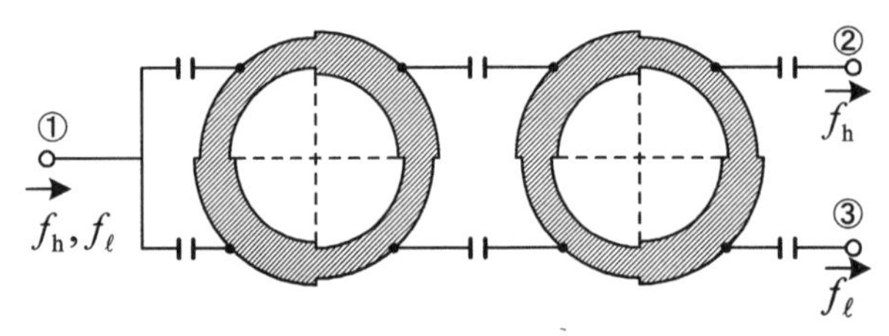

Fig. 5.8. Schematic configuration of diplexer using ring type dual-mode SIRs

$$\begin{aligned}\frac{\Delta f}{f_0} &= \frac{1}{f_0}\left|f_{01} - f_{02}\right| \\ &= \frac{4}{\pi}\left|\theta_{01} - \theta_{02}\right| \\ &= \frac{4}{\pi}\left|\tan^{-1}\sqrt{R_Z} - \tan^{-1}\sqrt{1/R_Z}\right| . \end{aligned} \tag{5.1}$$

Figure 5.7 illustrates the relationship between R_Z and frequency separation. By choosing $R_Z = 0.5$, a $\Delta f / f_0$ of 0.43 can be obtained, implying that a diplexer with wide pass-band separation can be designed. A multistage diplexer utilizing the above characteristics can be realized by a cascaded connection of ring resonators with two bandpass filters designed to satisfy the specifications of each passband as shown in Fig. 5.8.

5.3 Application of λ_g-Type SIR as Two-Port Devices

5.3.1 Coupling Means for Orthogonal Resonant Modes

As previously described, for a two-port application, the two orthogonal modes existing within a ring resonator are used as coupled modes instead of inde-

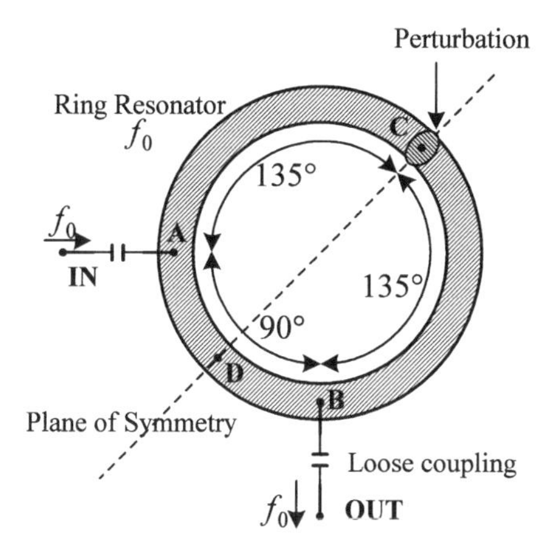

Fig. 5.9. Basic structure of a dual-mode filter using a stripline ring resonator

pendent use. Thus, the most typical application of two-port devices are band pass filters. Figure 5.9 illustrates the basic structure of a dual-mode filter using one-wavelength ring resonators. Input and output ports are spatially separated at 90° intervals, and an adequate perturbation or discontinuity is positioned at an equal distance from the input and output ports (point C or D in the figure). The behavior of the combination of two orthogonal resonance modes can qualitatively be explained by applying the traveling wave theory to this figure. For simplicity, we assume a structure where the resonator is excited by electric coupling, and the input/output ports are capacitive coupled as shown in the figure. We first consider a case eliminating the perturbation or discontinuity within the resonator. The incident wave excited at the input port generates a strong electric field at point A due to electric coupling. The electromagnetic waves due to this electric field propagate clockwise and counterclockwise reaching point B at a reverse phase, where the phase of the clockwise wave is 270° while the counterclockwise wave is 90°. Thus, as explained in Sect. 5.1, the electric field amplitude at point B due to these waves becomes zero, and no response is generated at the output port.

Bearing this behavior in mind, we next position a discontinuity such as stub or notch at point C. As with the previous case, a strong electric field is generated at point A by exciting the input port, and thus two travelling waves are excited. The counterclockwise wave reaches point B at a 90° phase shift. Further propagating to point C at a 135° phase shift, a portion of this wave is reflected at in-phase or reverse phase due to the discontinuity in the transmission line of the resonator. Considering the reflected wave as in-phase, this wave propagates to point B while encountering a further phase shift of 135°. The total phase shift from point A becomes 360°, and thus the electric field magnitude of this reflected wave attains its maximum value at this point. This allows the wave to propagate through the output port by electric coupling. Similarly, the reflected wave of the clockwise travelling wave due

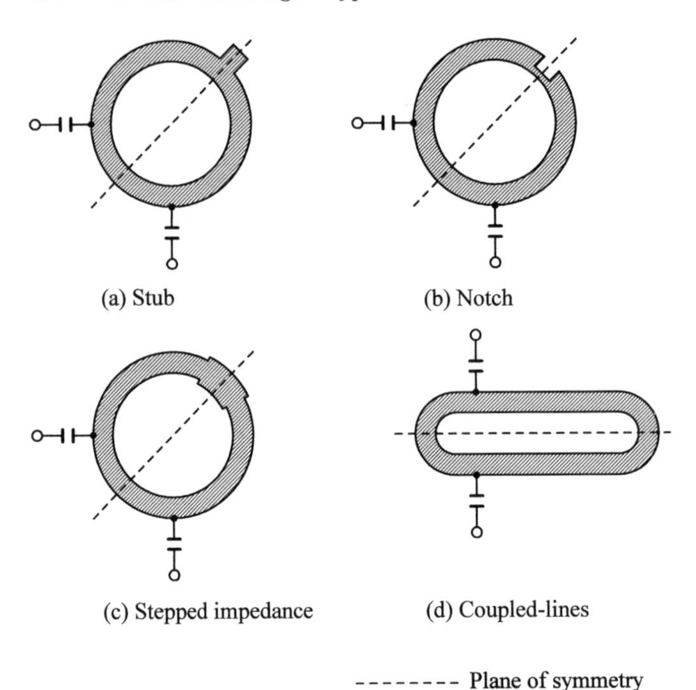

Fig. 5.10. Structural variations of a dual-mode stripline ring resonator filter

to the discontinuity can also propagate through the output port. When the reflected wave is excited in this form, the spatial phase relationship between the incident and the reflected wave can be expressed as mode#1 and mode#2 in Fig. 5.1. This implies the simultaneous existence of two orthogonal modes, claiming that the two orthogonal modes can be coupled by the presence of discontinuity in the ring resonator.

The coupling strength between the two modes will depend on the amplitude of the reflected wave, and thus is determined by the structure of the discontinuity or perturbation. The above discussion illustrates the nature of coupled mode behavior. General conditions [5] realizing a dual-mode filter using one-wavelength ring resonator are as follows:

1) The input and output ports should be spatially separated at 90° intervals.
2) A discontinuity or an alternative means of generating a reflected wave against the incident wave should exist within the resonator.
3) Symmetrical circuit geometry should be applied.

Figure 5.10 illustrates structural examples of a typical dual-mode filter satisfying these conditions. The stepped impedance structure shown in (c) has two advantages and is often applied to dual-mode filters. The first is a highly flexible design obtained by controlling the two parameters of impedance ratio and step length. The second is the step discontinuity structure itself, which allows for a negligible center frequency shift. The following discussions focus on this stepped impedance structure as a discontinuity element, presenting the analysis method of dual mode coupling and practical application examples of the ring resonator.

5.3.2 Analysis of Coupling Between Orthogonal Resonant Modes

As previously described, a dual-mode filter based on a ring resonator structure is considered a symmetrical two-port circuit, and thus can be analyzed as a one-port circuit by applying a conventional even- and odd-mode excitation method. Even-mode excitation, where two in-phase signals of equal amplitude are simultaneously applied to the input and output ports, satisfies open-circuited conditions at the symmetrical plane of the circuit. Thus the circuit can be divided into two identical sub-circuits at the symmetrical plane, and analysis is based on either sub-circuit by applying an open-circuited condition to the divided plane. Similarly for odd-mode excitation, where signals of reverse-phase are applied to the input and output ports, an identical circuit satisfying short-circuited conditions at the divided plane is considered. The resonance conditions for each mode are analyzed from the above sub-circuit, and the derived resonance frequencies are used to obtain the coupling between the two orthogonal resonance modes.

The coupling coefficient k between the orthogonal resonance modes can be expressed in the following form [6] using the even- and odd-mode resonance frequencies, respectively expressed as f_{even} and f_{odd}.

$$k = \frac{2\,|f_{\text{even}} - f_{\text{odd}}|}{f_{\text{even}} + f_{\text{odd}}}. \tag{5.2}$$

Figure 5.11 illustrates equivalent circuit expressions of the SIR ring resonator excited by even- and odd-mode conditions, where the step junction is centered on point C. The circuit becomes open-circuited at points C and D for even-mode excitation, while for odd-mode excitation the same points obtain short-circuited conditions. As previously discussed, the discontinuity within the

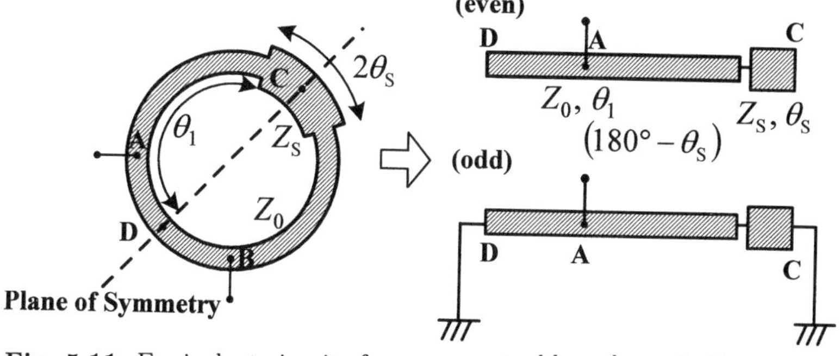

Fig. 5.11. Equivalent circuits for even- and odd-mode excitation

circuit can be centered on point D, and the analysis method described below can still be applied. Here we focus on point C as the center of discontinuity, for which a step junction is applied in this case.

The characteristic impedances of the ring resonator and the step portion are expressed as Z_r and Z_S, respectively. The electrical length of the step portion at the center frequency f_0 (resonance frequency of a UIR without a step junction) is $2\theta_S$, and $\theta_1 = \pi - \theta_S$. In addition, the even- and odd-mode resonance frequencies normalized by center frequency f_0 are expressed as f_{ne} $(= f_{even}/f_0)$ and f_{no} $(= f_{odd}/f_0)$, and impedance ratio $R_Z = Z_S/Z_r$. Using the above parameters, the even- and odd-mode resonance frequencies are derived as the following equations.

$$\text{For even-mode}: R_Z \tan(\theta_1 f_{ne}) + \tan(\theta_S f_{ne}) = 0. \quad (5.3)$$

$$\text{For odd-mode}: \tan(\theta_1 f_{no}) + R_Z \tan(\theta_S f_{no}) = 0. \quad (5.4)$$

The following discussions focus on the attenuation poles which are frequently generated in the attenuation region near pass-band. These poles are caused by the perturbation within the ring resonator, and are extremely important from a practical point of view due to a capability of improving attenuation characteristics in the stop-band. For a symmetrical circuit such as a dual-mode resonator, the attenuation poles emerge under a frequency condition where the intrinsic impedance of both modes becomes equal. Under these conditions, the currents due to the odd- and even-mode encounter a reverse phase at the output port, thus canceling each other out. In the case of a one-wavelength ring resonator, the attenuation poles can be analyzed in the following manner.

Considering the two propagation paths between the input and output ports shown Fig. 5.12a, the Y matrix of the corresponding circuits are expressed as Y' and Y''. The dual-mode filter is equivalent to a parallel connection of these two circuits, as shown in Fig. 5.12b. The output voltage becomes zero at a frequency inducing the attenuation pole, because the input signal applied to terminal $1 - 1'$ does not appear at output terminal $2 - 2'$. The output current I_2 also becomes zero at this pole frequency, and thus the following condition is derived:

$$I_2' = -I_2''.$$

Consequently, the circuit condition generating the attenuation poles is given as,

$$Y_{21}' = -Y_{21}''. \quad (5.5)$$

Thus, the attenuation pole frequency can be obtained by solving the above equation. However, (5.5) is valid only when the attenuation poles actually exist, and depending on the discontinuity structure, conditions where they do not appear could also exist. Furthermore, this structure can be altered to generate attenuation poles at both upper and lower sides of the pass-band, or at only either side.

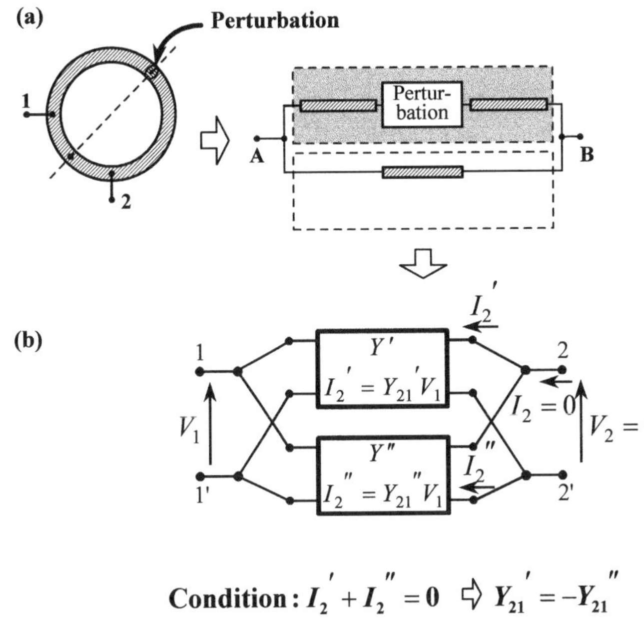

Fig. 5.12. Analysis method of attenuation poles in a dual-mode filter. (**a**) Equivalent expression with two propagation paths. (**b**) Corresponding circuit expression using Y-matrices

Figure 5.13 illustrates the relationship between step length and f_{ne}, f_{no}, and also between step length and normalized pole frequency, for $R_Z = 0.5$. The coupling coefficient can be obtained from f_{ne} and f_{no} by applying (5.2). The figure suggests that the frequency span between f_{ne} and f_{no} enlarges in accordance with step length.

Figure 5.14 shows calculated values of even- and odd-mode resonance frequencies and attenuation pole frequency as a function of impedance ratio R_Z $(= Z_{\mathrm{S}}/Z_{\mathrm{r}})$, for a constant step length of 5°. Though partly due to a short step length, the results imply that a large coupling coefficient cannot be achieved by simply changing R_Z.

$R_Z = 1$ represents a uniform impedance resonator (UIR), for which there is no coupling between the two orthogonal modes due to $f_{\mathrm{ne}} = f_{\mathrm{no}}$. Under the condition of $R_Z > 1$, the figure suggests that the attenuation poles will not be generated, due to a SIR structure disqualifying (5.5). Thus, we understand that the attenuation poles are dependent on the perturbation structure, and cases where they do not exist can be assumed.

Figure 5.15 shows an experimentally observed example of even- and odd-mode resonance and the attenuation poles. Measurements were based on conditions of $f_0 = 1.9\,\mathrm{GHz}$, $Z_{\mathrm{r}} = 50\,\Omega$, $Z_{\mathrm{S}} = 25\,\Omega\,(R_Z = 0.5)$, and $2\theta_{\mathrm{S}} = 20°$. The coupling strength between input and output port are loosely set to avoid any interference to the resonance characteristics. Calculated and

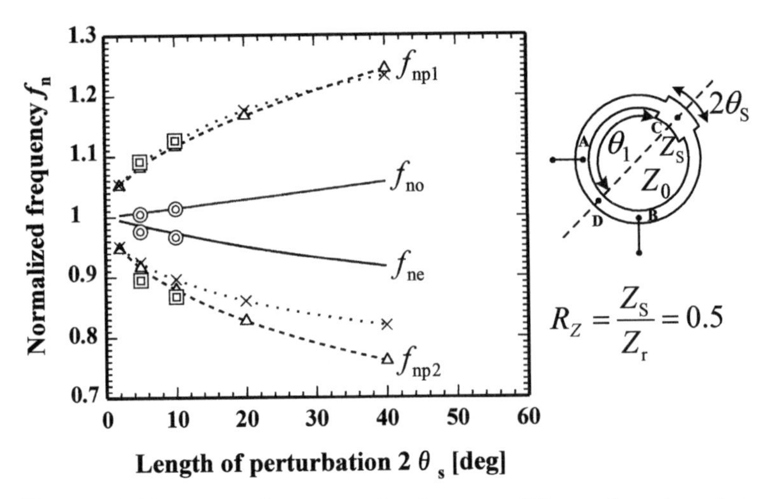

Fig. 5.13. Resonance frequency of a ring-type SIR as a function of step length $2\theta_S$

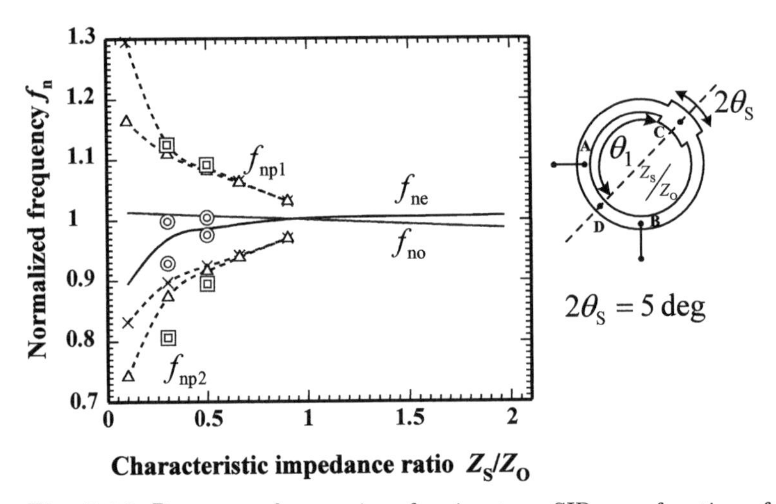

Fig. 5.14. Resonance frequencies of a ring type SIR as a function of impedance ratio R_Z

measured results, including the coupling coefficient between the orthogonal modes, are summarized in Table 5.1. With the exception of a slightly low value for measured frequency, the results show good agreement, thus proving the validity of the analysis method described.

The above discussion is based on a case where the perturbation or discontinuity within the transmission line is centered on point C in Fig. 5.11. As previously described, this analysis method can be applied to a case where the perturbation is centered on point D, and similar properties can be obtained through analysis.

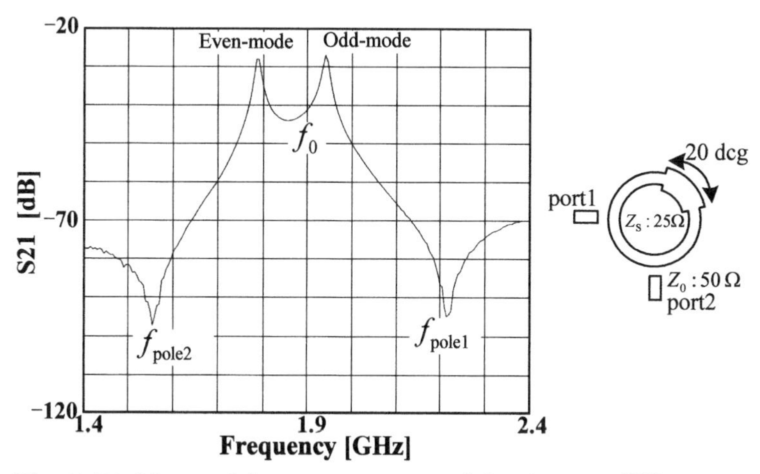

Fig. 5.15. Measured frequency response of the ring type SIR

Table 5.1. Comparison of calculated and measured data

	Calculated	Measured
f_{ne}	0.950	0.945
f_{no}	1.028	1.025
k	0.040	0.041
f_{np1}	1.167	1.163
f_{np2}	0.827	0.817

5.3.3 Application to Filters

The basic structure of a multistage bandpass filter using ring-shaped SIR is a cascaded connection of dual mode filters consisting of one-ring resonators. Here we discuss the design technique for a basic dual-mode filter applying SIR. The difference between a conventional resonator-coupled filter and an orthogonal resonance mode filter is that in the former case, the resonator and coupling circuits can be divided, thus allowing discussions to be held separately, while for the latter this cannot be done. An example in the case of a dual mode filter is that the attenuation characteristics differ by the coupling circuit structure, namely the discontinuity structure and its location, even if coupling strength is kept constant. This implies that even for a fixed specification of pass-band bandwidth, the attenuation characteristics will change in accordance with the discontinuity structure. Our discussions focus on a filter design method in which pass-band characteristics are regarded prior to attenuation performance, based on the conditions of a given pass-band bandwidth. An approximate design method is applied by assuming a two-stage BPF where capacitive coupling is adopted for input and output coupling.

The center frequency f_0, relative bandwidth w, and element value g_0, g_1, g_2, g_3 of two-stage BPF are assumed given from the design specifications. As described in the appendix, the interstage coupling coefficient k_{12} is expressed as,

$$k_{12} = \frac{w}{\sqrt{g_1 g_2}}$$

When applying SIR as a resonator element, it is required to determine the step structure according to the discussion in Sect. 5.3.2. Input and output coupling capacitor C_S is given from the equation derived for capacitor coupled BPF in Sect. 3.2.1.

$$C_S = \frac{J_{01}}{\omega_0 \sqrt{1 - (J_{01}/G_S)^2}},$$

$$J_{01} = \sqrt{\frac{G_S b_r w}{g_0 g_1}},$$

G_S : Source conductance

b_r : Resonator slope parameter.

In addition, the slope parameter b_r can be obtained under the assumption of a one-wavelength resonator as,

$$b_r \cong \pi / Z_r = \pi Y_r .$$

Due to a dependency on the discontinuity structure and location, attenuation performance must be confirmed by the direct calculation of transmission response. Although an approximate design can be obtained from the above discussion, these calculations are merely assumptions due to the difficulty of separating the resonator and coupling circuits, and thus center frequency shift due to the coupling circuits yield adjustment of the physical dimension of the ring resonator. However, as illustrated above, the structural simplicity of the dual-mode filter enables easy calculation of an approximate transmission response, and by applying these results as initial values, rigorous design parameters can be obtained through CAD optimization.

Based on the above design method, three experimental dual-mode filters with a center frequency f_0 of 1.9 GHz and relative bandwidth of 0.5%, 1.4%, and 2.0%, were designed and fabricated using ring-type SIR. The impedance ratio R_Z was chosen as 0.8 ($Z_r = 50\,\Omega$, $Z_S = 40\,\Omega$), and the coupling length was controlled by adjusting the length of transmission line Z_S. The design parameters are summarized in Table 5.2. Figure 5.16 shows a photograph of the experimental filters, fabricated on a substrate measuring 27 mm×27 mm with a relative dielectric constant $\varepsilon_r = 10.5$ and thickness H=1.27 mm. Lumped-element chip capacitors are adopted as input and output coupling capacitors. Calculated attenuation and return loss characteristics are illustrated in Fig. 5.17, while actual measured results are shown in Fig. 5.18.

Table 5.2. Dual-mode filter design parameters
Center frequency $f_0 = 1.9\,\text{GHz}$
Impedance Ratio $R_Z = 0.8\,(Z_r = 50\,\Omega, Z_S = 40\,\Omega)$

Filter Type	I	II	III
Relative Bandwidth w	0.5%	1.4%	2.0%
Step Length $2\theta_S$	8°	18°	28°
Coupling Capacitance C_S	0.28 pF	0.44 pF	0.54 pF

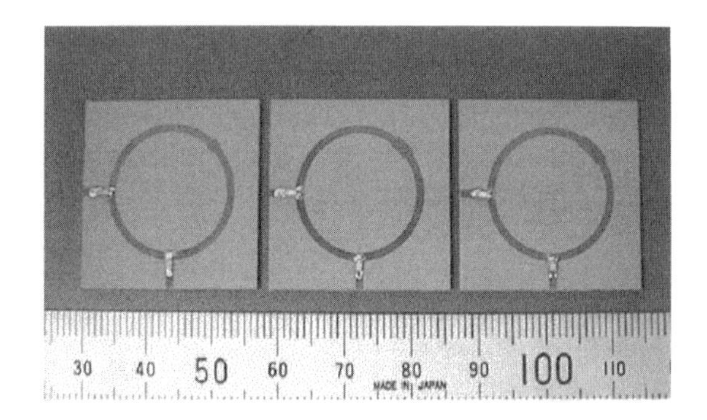

Fig. 5.16. Photograph of the experimental dual-mode filters

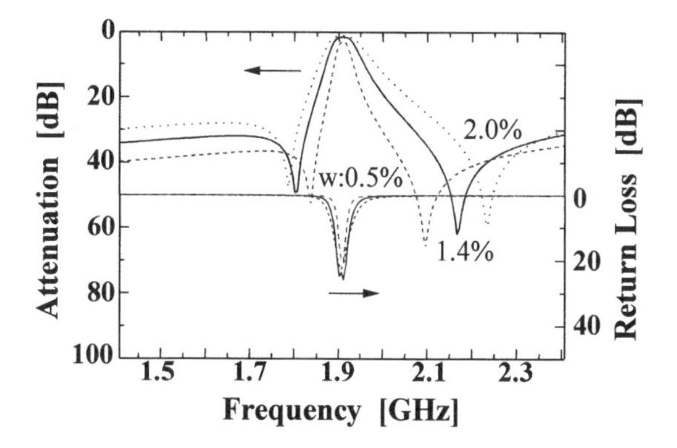

Fig. 5.17. Calculated responses of the experimental filters

A good agreement is seen between design and actual results, thus demonstrating the validity of the design method described. In addition, the attenuation poles appear in both the upper and lower sides of the pass-band, giving a steep gradient of attenuation in spite of a two-stage BPF structure. This circuit behavior is illustrated as one of the features of a dual-mode filter using ring-type SIR.

Low radiation properties are another feature of the ring resonator worthy of note. Radiation loss generally shows a remarkable increase in the millime-

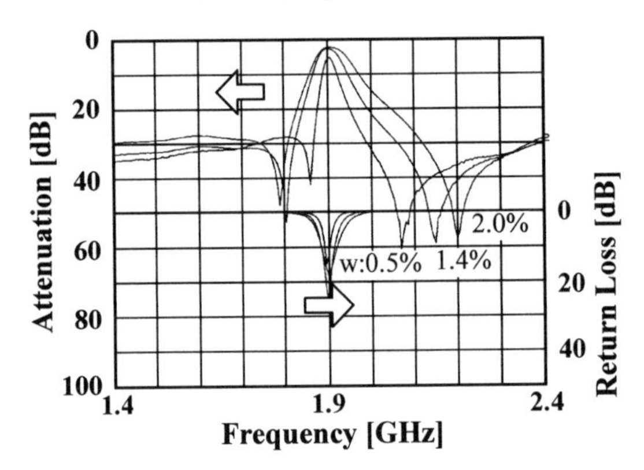

Fig. 5.18. Measured responses of the experimental filters

ter wave region, and thus it becomes difficult to adopt the $\lambda_g/2$ type uniform impedance resonator with open-circuited ends, which is conventionally used in the microwave region. In addition, the ring resonator gives more allowance to physical dimensions because of its one-wavelength resonator structure, resulting in a practical advantage of easy manufacturing. These features create high hopes for their application to millimeter wave integrated circuits.

Described here is a dual mode filter for millimeter wave ICs based on a Si substrate. In an attempt to realize high-performance low-cost millimeter wave ICs, a new manufacturing process has been developed, in which active circuits are formed on compound semiconductor chips such as GaAs, and mounted on low-cost Si substrates by flip-chip mounting technology. Experimental examples applying this new technology, such as a receiver front-end IC, have been reported.

Transmission lines such as micro-striplines formed on a conventional Si substrate possess a shortcoming of large transmission loss due to the dielectric loss of the substrate. Thus, transmission-line resonator filters constructed on such Si substrates could not realize the required performance for practical use.

To overcome this problem, a transmission-line structure applying micromachining technology has been proposed, and an experimental filter based on this technology has been fabricated to verify its availability [7]. Figure 5.19 compares the structures of a conventional micro-stripline (MSL) to an inverted micro-stripline (IMSL) realized by the above technology.

The electrical field of an ordinary microstripline is concentrated in the Si substrate, thus causing a large dielectric loss. In the case of an inverted microstripline, the electric field is focused on the air cavity where Si material is removed by micromachining technology, thus resulting in an extremely small dielectric loss. Figure 5.20 compares calculated transmission losses of the micro-stripline and inverted micro-stripline. The figure suggests that the influence due to the $\tan\delta_d$ of the substrate becomes smaller for an inverted micro-stripline, reducing loss to 1/3–1/2 as compared to a conventional micro-stripline. In addition, as is apparent from Fig. 5.19, the skin effect enables the

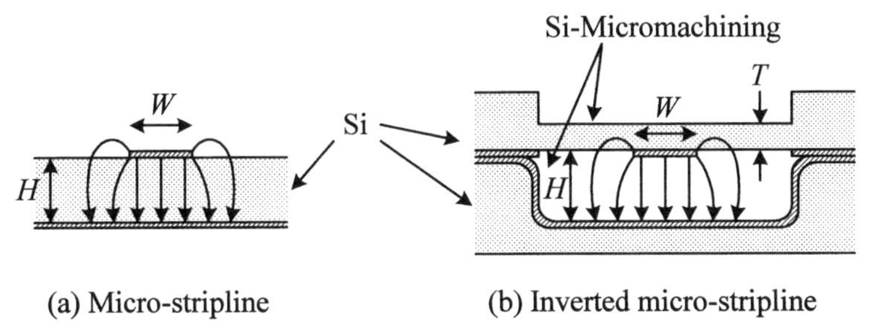

Fig. 5.19. Structure of micro-striplines fabricated on Si-substrate

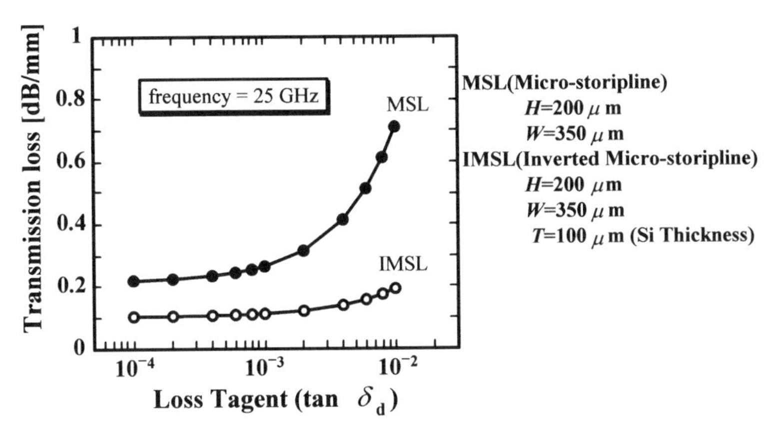

Fig. 5.20. Calculated transmission losses of micro-stripline (MSL) and inverted micro-stripline (IMSL)

current to flow on the surface of the metal conductor contacting the air gap, while only a minute portion of it will flow on the boundary surface of the Si substrate. This enables the inverted micro-stripline structure to further reduce conductor losses.

We next exemplify the application of a dual mode ring-type SIR filter based on this inverted micro-stripline structure, intended for the receiving filter of a 30 GHz band wireless communication terminal. Design specifications of this experimental filter are as follows.

Center frequency	: $f_0 = 31.5\,\text{GHz}$
Pass-band bandwidth	: $W > 1.0\,\text{GHz}$
Attenuation	: $L_S > 15\,\text{dB}$ at 35 GHz
Pass-band insertion loss	: $L_0 < 1.5\,\text{dB}$

The coupling coefficient of orthogonal modes within the ring resonator, which satisfies specifications, is chosen as $k = 0.047$, and thus the resonator parameters are determined as follows;

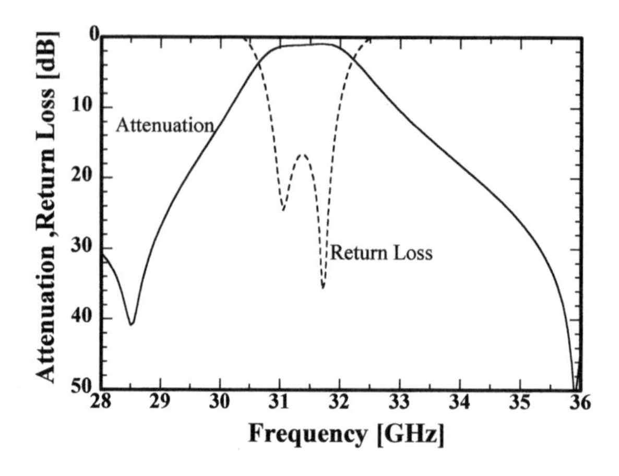

Fig. 5.21. Calculated responses of the experimental ring type SIR dual-mode filter

$$Z_r = 90\,\Omega,$$

$$Z_S = 70\,\Omega, \quad 2\theta_S = 36^\circ.$$

As a result, the attenuation poles are calculated as

$$f_{p1} = 28.5\,\text{GHz}$$

$$f_{p2} = 36.0\,\text{GHz}.$$

Consequently, an attenuation of 24 dB, which meets specifications, is obtained at 35 GHz. Figure 5.21 shows calculated transmission characteristics of the filter realized by these design parameters. The structural parameters of the applied inverted micro-stripline are chosen as follows;

Line width W_r	$= 300\,\mu\text{m}$ ($90\,\Omega$)
W_S	$= 500\,\mu\text{m}$ ($70\,\Omega$)
Center conductor height H	$= 200\,\mu\text{m}$
Thickness of Si substrate T	$= 500\,\mu\text{m}$.

A calculated effective dielectric constant ε_{eff} of 1.11 and an unloaded-Q of approximately 300 was obtained for this structure.

Figure 5.22 illustrates the filter circuit pattern, while measured frequency response is shown in Fig. 5.23. Measured data showed good agreement with designed values for frequency, while the observed pass-band bandwidth was slightly wider than expected. Connections between the two substrates of different structure are assumed to be responsible for such deviation, along with the manufacturing accuracy of air gap height H. Even so, design tolerance allowed attenuation characteristics to meet specifications. In addition, a measured unloaded-Q value of approximately 300, which closely corresponds to the theoretical value, was obtained, and consequently the measured pass-band insertion loss of 0.7 dB also satisfied target specifications. These results obtained satisfy actual characteristics required for the receiving filter of practical radio systems, and thus the development of such millimeter wave

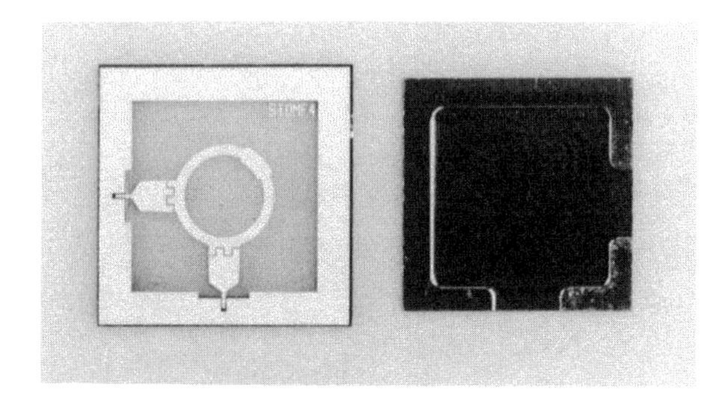

Fig. 5.22. Photograph of the experimental dual-mode filter

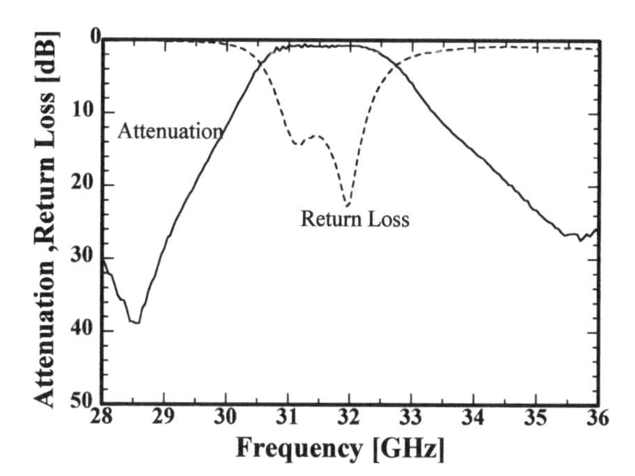

Fig. 5.23. Measured response of the experimental dual-mode filter

ICs based on Si substrates are expected to progress in the field of wireless communications. Moreover, the introduction of micromachining technology has realized practical low-loss transmission lines based on a Si substrate, and with this technology we forecast accelerated development of high performance cost-effective millimeter wave ICs including built-in antenna structures. The dual-mode filter using ring-type SIR possesses good affinity with such millimeter wave ICs, and thus is considered a promising candidate for filters of future application.

6. Expanded Concept and Technological Trends in SIR

In the previous parts we considered the SIR as a distributed transmission-line resonator possessing a step-junction structure. When focusing on broad electrical characteristics to classify single-dimensional transmission-line resonators into homogeneous and inhomogeneous impedance resonators, the SIR can be categorized as an inhomogeneous impedance resonator. Thus, by expanding the basic concept of the SIR defined in the previous chapters, an expanded analysis method can be applied to inhomogeneous impedance resonators, such as tapered-line resonators possessing a continuous change in characteristic impedance. Although in the former discussions the step change in characteristic impedance is realized by a geometrical change in the structure of the transmission line, an equivalent impedance step can be achieved by altering the material applied to the transmission line, or by a combination of both structural and material changes.

In this chapter, we illustrate the possibilities of applying an expanded SIR concept to analyze and design various resonator structures including composite material, multistep, tapered-line, and folded-line resonators. Discussions further extend to the practical applications of such resonator structures, finishing with some interesting topics from a practical point of view, concerning the future technological trends of the SIR.

6.1 SIR Composed of Composite Materials

6.1.1 Combination of Magnetic and Dielectric Materials

Previous discussion presumed a homogeneous transmission-line medium, where the impedance step of the SIR is generated by a geometrical change in a transmission-line structure. However, an impedance step can be achieved while maintaining physical dimensions of the transmission line, simply by changing the material of the transmission line, or for a larger impedance change, by changing both material and structure.

Figure 6.1 illustrates a $\lambda_g/4$ type coaxial SIR structure based on composite materials [1,2]. Previous discussions suggest that the diameter ratio b/a between inner and outer conductors should be selected as 3.6 to maximize the unloaded-Q value of a coaxial resonator. By applying this design criteria,

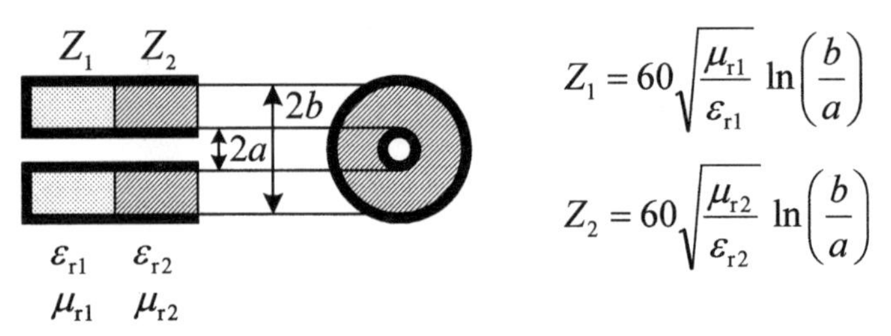

Fig. 6.1. Structure of $\lambda_g/4$ type coaxial SIR using composite materials

this hybrid SIR structure is expected to realize further size reduction and an even higher unloaded-Q. ε_{r1} and ε_{r2} in the figure indicate relative dielectric constant (relative permittivity), while μ_{r1} and μ_{r2} indicate relative permeability. The characteristic impedance of a coaxial transmission-line is given by,

$$Z = 60\sqrt{\frac{\mu_r}{\varepsilon_r}}\ln(b/a) \cong 77\sqrt{\frac{\mu_r}{\varepsilon_r}}.$$

The impedance ratio R_Z can be obtained as,

$$R_Z = \frac{Z_2}{Z_1} = \sqrt{\frac{\mu_{r2}}{\varepsilon_{r2}}}\Big/\sqrt{\frac{\mu_{r1}}{\varepsilon_{r2}}} = \sqrt{\left(\frac{\mu_{r2}}{\mu_{r1}}\right)\left(\frac{\varepsilon_{r1}}{\varepsilon_{r2}}\right)} \tag{6.1}$$

The wavelength reduction factor is given by $1/\sqrt{\varepsilon_r\mu_r}$, and thus conditions where $\mu_{r2} < \mu_{r1}$ and $\varepsilon_{r1} < \varepsilon_{r2}$ while obtaining a maximum μ_r, ε_r become indispensable for miniaturization. This implies that for an ideal configuration, it is desirable to use a magnetic material for the Z_1 section and a dielectric materials for the Z_2 section. However, magnetic material possessing excellent characteristics in the microwave region are scarce, and thus a SIR structure based on several dielectric materials of varied dielectric constant has been proposed and experimentally studied.

6.1.2 Coaxial SIR Partially Loaded with Dielectric Material

Conditions for miniaturization of SIR based on composite materials have been discussed in the previous section. As a practical solution, here we consider the structure of a SIR consisting of a combination of dielectric materials. One reported example [3–5] employs air for the Z_1 section and a material of high dielectric constant for the Z_2 section, this corresponding to the conditions of $\mu_{r1} = \mu_{r2} = \mu_{r0} = 1$ and $\varepsilon_{r2} > \varepsilon_{r1} = \varepsilon_{r0} = 1$. Figure 6.2 illustrates the actual SIR structure. The resonance conditions are the same as a basic SIR, given as,

$$\tan\theta_1 \tan\theta_2 = R_Z.$$

Total resonator length and partial lengths ℓ_1 and ℓ_2, normalized by the $\lambda_g/4$ length in free space, are, respectively, expressed as L_{nt}, L_{n1}, and L_{n2}. The

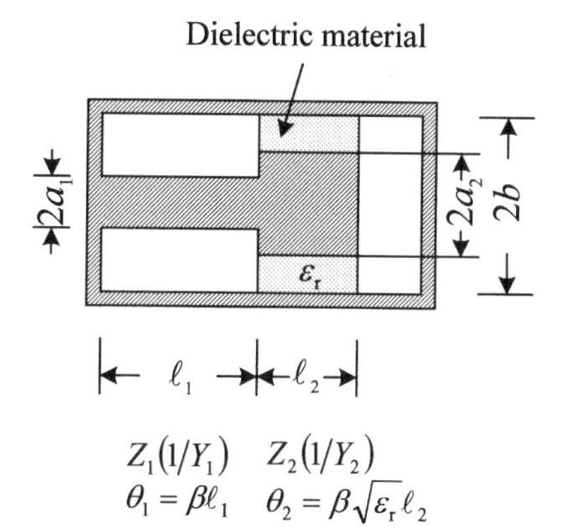

Fig. 6.2. $\lambda_g/4$ type coaxial SIR partially loaded with high-dielectric-constant ceramic

relationship between L_{nt} and L_{n1} is obtained from the resonance condition as,

$$L_{nt} = \frac{2}{\pi\sqrt{\varepsilon_r}}\left[\tan^{-1}\left(\frac{T}{1-R_Z}\right) + \frac{\pi L_{n1}}{2}(\sqrt{\varepsilon_r}-1)\right], \tag{6.2}$$
$$\text{where} \quad T = \tan(\pi L_{n1}/2) + R_Z/\tan(\pi L_{n1}/2).$$

Figure 6.3 shows the calculated results for a case assuming $\varepsilon_r = 85$. These results illustrate the basic features of a SIR partially loaded with dielectric material, which are summarized as follows.

1) The total resonator length is determined by ε_r and R_Z.
2) When $R_Z < 1/\sqrt{\varepsilon_r}$, the total resonator length possesses a minimum value which is substantially shorter than that of a dielectric loaded UIR.
3) When $R_Z \geq 1/\sqrt{\varepsilon_r}$, the total resonator length possesses no minimum value, and is longer than that of a dielectric loaded UIR.

For the unloaded-Q value, the analysis method described in Chap. 3 can be applied, and analytical results have proved similar basic trends [4]. Although filters and duplexers based on such SIR structure have been reported [3,6], these examples show little advantage concerning miniaturization when compared to similar circuits based on double coaxial dielectric SIR. Yet for high power applications in frequency regions below VHF, difficulties are encountered in the fabrication of available ceramic materials, and thus partially loaded SIR structures, which consume less material mass, are considered a highly available solution due to low production cost and light weight. The concept of partially loading dielectric material to a coaxial structure can effectively be applied to a low-pass filter (LPF) based on a stepped impedance structure. A distributed-element type transmission-line LPF is composed of alternately connected transmission lines of high and low impedance, where miniaturization can effectively be achieved by applying dielectric material to the low impedance lines. Figure 6.4 illustrates a VHF band 5-stage coaxial LPF structure based on partially loaded dielectric ceramics, while Fig. 6.5

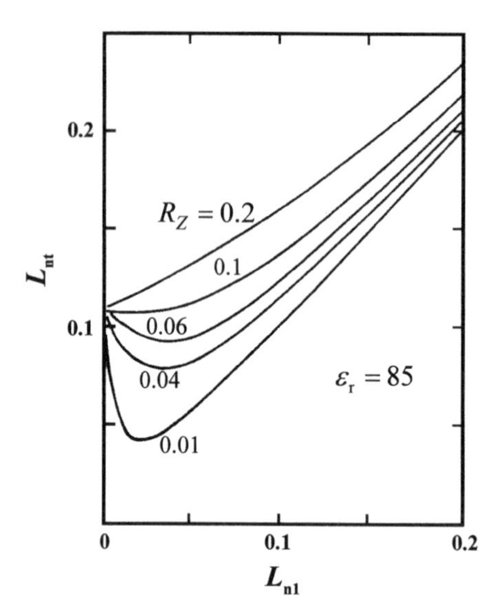

Fig. 6.3. Relationship between L_{nt} and L_{n1} of SIR partially loaded with dielectric material

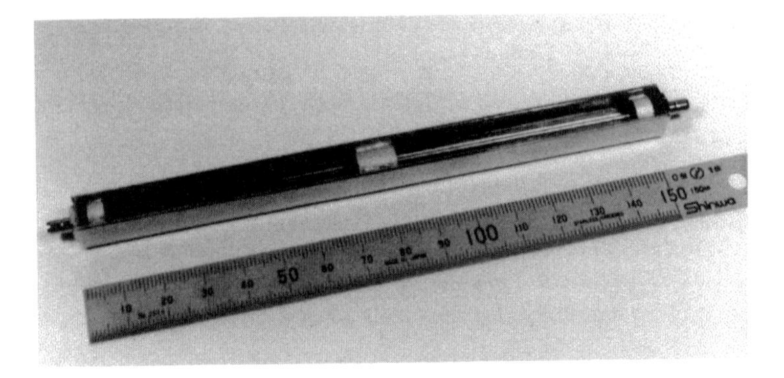

Fig. 6.4. Photograph of the experimental LPF possessing stepped-impedance structure

shows actual electrical performance. Transmission line LPFs are known for their good electrical properties, but due to a handicap of large size, they are usually applied in frequency regions above the UHF band. However, by introducing a stepped-impedance structure employing dielectric materials, we understand that withstand practical application even in the VHF band. The relative dielectric constant of the ceramics used in this example is 36, and overall filter size measures $150\,\mathrm{mm}(L)\times 10\,\mathrm{mm}(W)\times 10\,\mathrm{mm}(H)$.

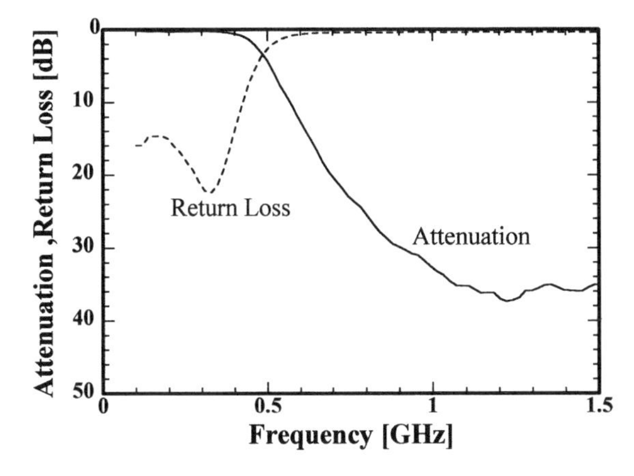

Fig. 6.5. Measured frequency responses of the experimental LPF

6.2 Multistep SIR and Tapered-Line Resonators

Discussions up to this point treat SIR as a resonator structure possessing two transmission lines of different impedance. Theoretically speaking, however, there is no necessity to restrict the transmission line impedance within two levels, and this structure is adopted only for design convenience. Consequently, a discussion based on the extended definition of SIR requires consideration of a multistep resonator structure shown in Fig. 6.6.

By increasing the step number, the multistep resonator approaches a tapered-line resonator at extreme conditions. Despite a scientific interest as an object of analytical study, the multistep resonator seems to possess little availability for practical use when compared with conventional two-step SIR and tapered-line resonators. Yet, when considering the difficulty accompanying the direct analysis of a linear tapered-line resonator (LTLR), we find it extremely convenient to introduce an analysis method based on a multistep approximation. In the following section we describe the filter design technique and application examples based on this approximation method.

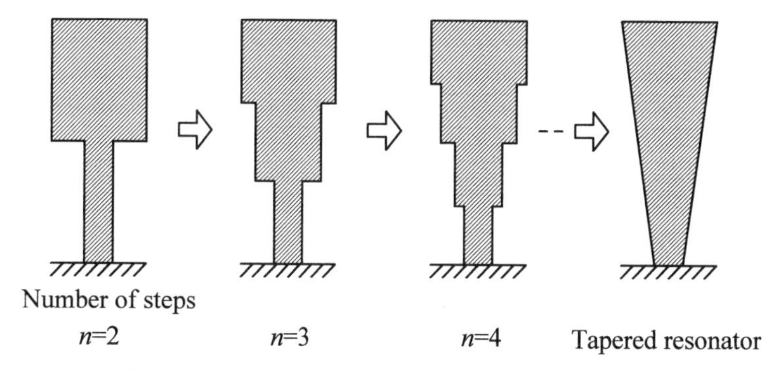

Fig. 6.6. Multistepped impedance resonators and linear tapered-line resonator

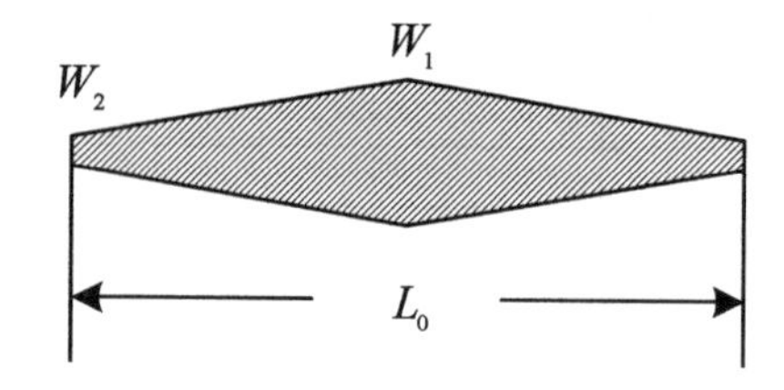

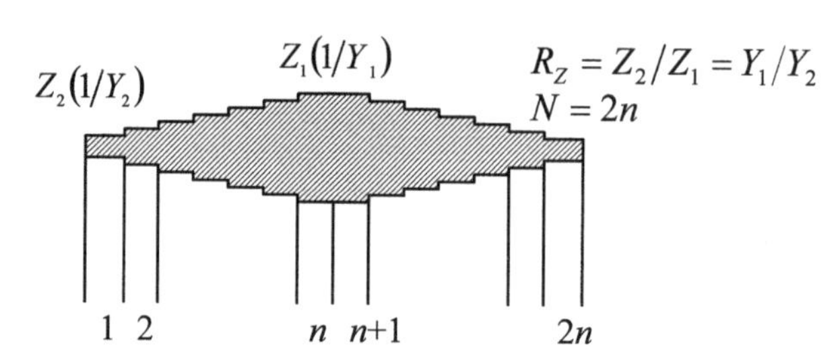

Fig. 6.7. Structure of the $\lambda_g/2$ type linear tapered-line resonator (LTLR) and its equivalent description using multistepped impedance resonator

Figure 6.7a illustrates the $\lambda_g/2$ type LTLR [1,7] with two open-circuited ends discussed in this section. This resonator structure is designed to reduce conductor loss by employing a wide line width near the center of the resonator where maximum current flows, while the open-ends, where no currents flow, are narrowed. In addition, the side edge coupling between resonators is easily realized by adopting a linear tapered structure, and this enables a strong coupling which realizes an availability for application to wide-band multistage BPF.

Figure 6.7b indicates an equivalent expression of the LTLR shown in (a) based on an N-stage multi-step SIR structure, where a good approximation is expected by increasing N. Resonator performance can be analyzed by obtaining the total F-matrix of the circuit from the F-matrices of the individual lines. This approximation method provides a direct analysis while saving calculation time as compared to an electromagnetic analysis of the resonator structure shown in (a). However, a quasi-TEM mode electromagnetic field distribution is assumed for this analysis method, thus possessing applicable limits which will be discussed later.

Characteristic impedance and line width at the open-ends and resonator center are expressed as Z_2, Z_1, and $W_2 W_1$, respectively, as shown in the figure. The impedance ratio R_Z is defined as $R_Z = Z_2/Z_1$. A micro-stripline structure is assumed, and edge effects at the open-ends, discontinuities in the step junctions, and changes of effective dielectric constant due to line width variation are all considered in the analysis. Figure 6.8 shows calculated total resonator length L_0 as a function of N for a substrate with a dielectric constant $\varepsilon_r = 10.5$ and thickness $H = 1.27$ mm, at resonance frequency $f_0 = 2.5$ GHz. Results illustrate a deviation of less than 0.5% for N >60, thus

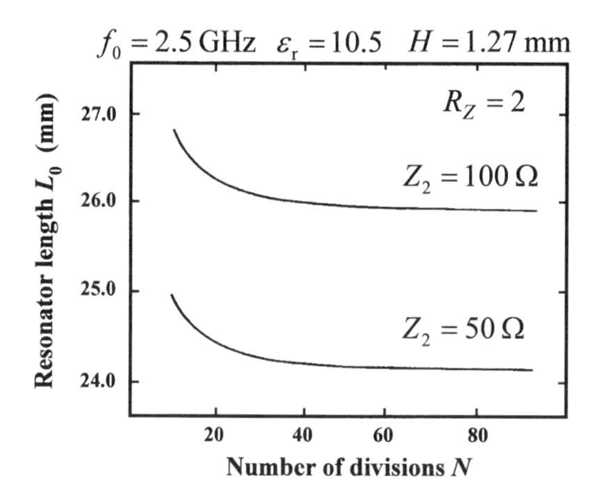

Fig. 6.8. Relationship between number of divisions N and resonator length L_0

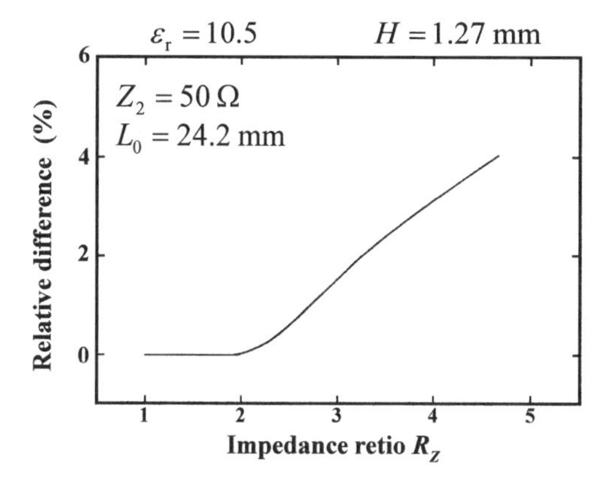

Fig. 6.9. Error estimation of resonance frequency of LTLR by multistep approximation

suggesting that a multistep resonator approximation employing a large N can effectively be applied to the analysis of linear tapered-line resonators. The threshold value of division number N allowing a deviation below 0.5% is independent of substrate dielectric constant and resonance frequency, and is mainly affected by the impedance ratio R_Z. In addition, for a large R_Z value, the resonator must be treated as a planar circuit, because the currents in the resonator show a two-dimensional distribution.

Figure 6.9 indicates the difference between calculated resonance frequency values obtained from rigorous electromagnetic analysis of a planar circuit and approximate estimations using a multistep one-dimensional model for LTLR. Results suggest that in order to achieve an error of less than 1%, the R_Z value must be below 2.5. These results illustrate the limitations of this multistep approximation method, while within these limits the method provides efficient results in an extremely short computation time as compared to conventional electromagnetic field analysis methods [7].

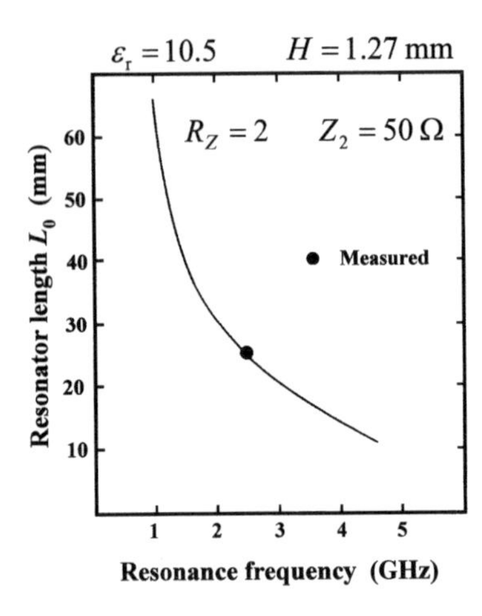

Fig. 6.10. Calculated results of the resonator length L_0

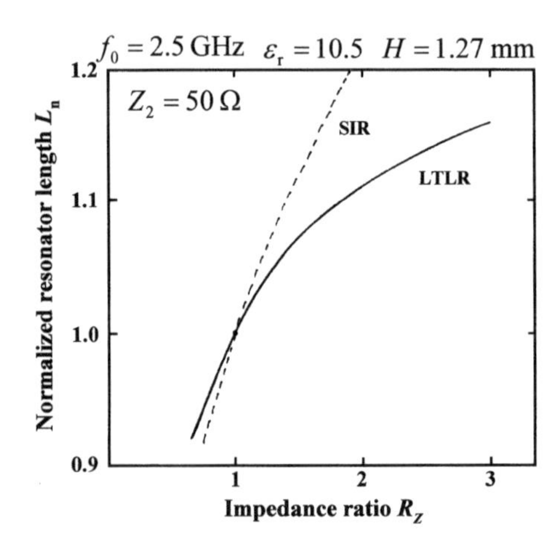

Fig. 6.11. Calculated results of the resonator length normalized by half-wavelength UIR

Considering the above results, the following discussions on LTLR are based on conditions of N of around 100 and $R_Z < 2.5$. Single resonator characteristics are first examined, followed by an analysis of resonator pairs to obtain resonator coupling properties required for filter design, and finally, these results are applied to the design and fabrication of an experimental filter. The applied substrate possesses a dielectric constant of $\varepsilon_r = 10.5$, and thickness H=1.27 mm.

The relationship between resonator length L_0 and resonance frequency f_0 for $R_Z = 2$ is shown in the Fig. 6.10, while the relationship between impedance ratio R_Z and normalized resonator length L_n for f_0=2.5 GHz is shown in Fig. 6.11. The dotted line in Fig. 6.11 indicates the normalized resonator length L_{n0} of a conventional two-step SIR ($\theta_1 = \theta_2$), and the results imply that the change of LTLR resonator length against R_Z is smaller than that of a two-step SIR.

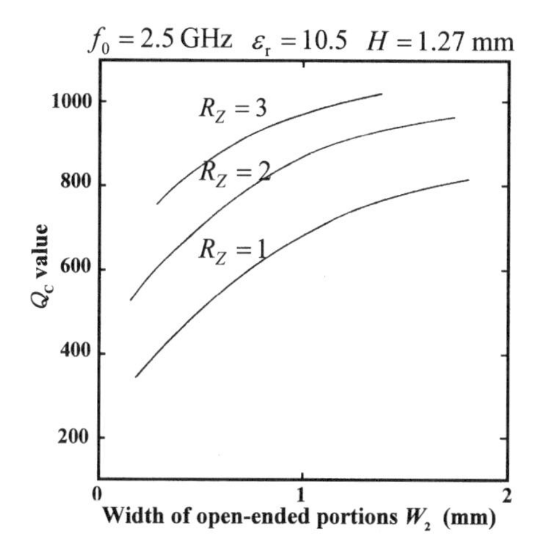

Fig. 6.12. Calculated results of Q_C of linear tapered-line resonator (LTLR)

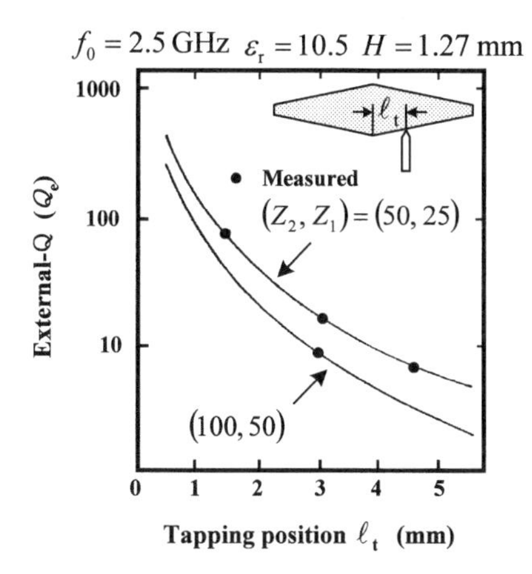

Fig. 6.13. Calculated results of external-Q (Q_e) by tapped-coupling

The calculated unloaded-Q values related to conductor losses (Q_C), obtained by the analysis method described in the appendix, are shown in Fig. 6.12. In this case also, the resonator is approximated by cascaded multi-step lines. For a constant W_2, Q_C increases in accordance with to impedance ratio R_Z, because the conductor losses are reduced by a wider transmission line width at the center of the resonator where a maximum current is observed. However, the direct influence of R_Z on total unloaded-Q is mitigated due to losses generated by the loss tangent of the substrate.

Figure. 6.13 illustrates calculated external-Q(Q_e) , which is required for determining the input/output circuit parameters in filter design. These results can also be obtained by applying the analysis method in the Appendix, where in this case a tapping method providing magnetic coupling is adopted as input/output circuits. Results are compared between two res-

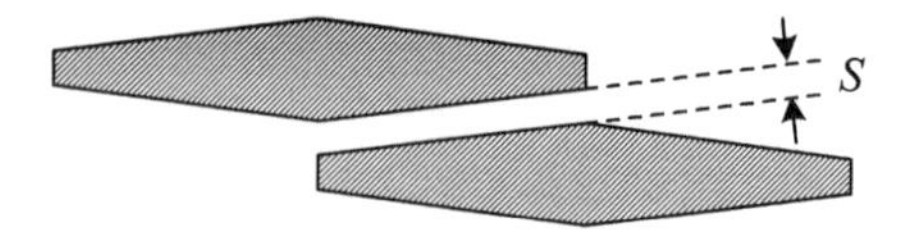

Fig. 6.14. Interstage coupling structure of linear tapered-line resonators (LTLRs)

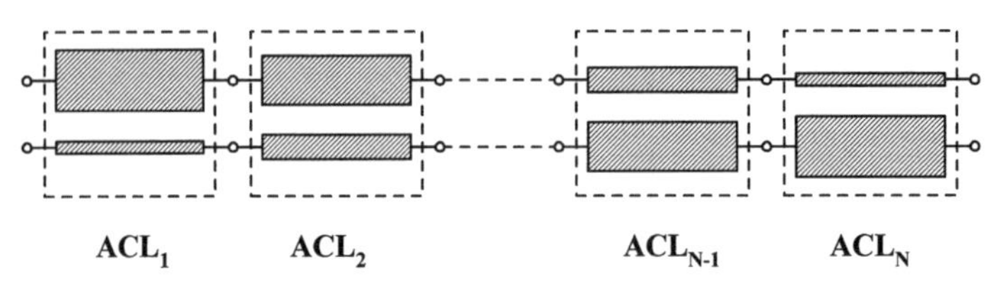

Fig. 6.15. An approximate description of the side-coupling section by asymmetrical coupled-lines

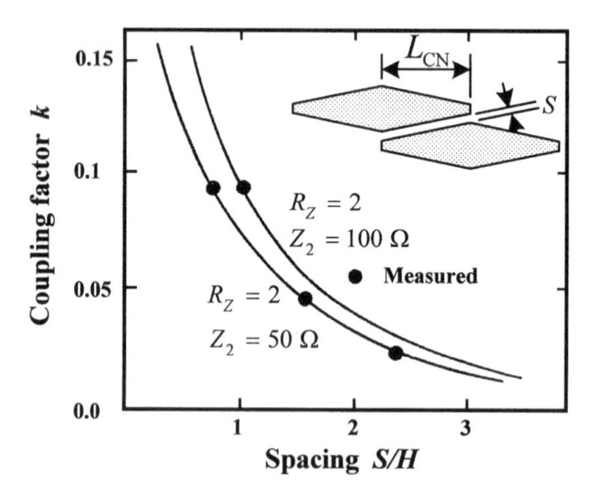

Fig. 6.16. Interstage coupling factor between linear tapered-line resonators (LTLRs)

onator structures both of $R_Z = 2.0$, where Q_e is influenced by transmission-line impedance due to tapping coupling configuration, resulting in different Q_e values for an identical tapping position ℓ_t.

For coupling between the resonators, a parallel arrangement of the resonators, as illustrated in Fig. 6.14, enables a side edge coupling structure which has an advantage for easy manufacturing. In this case also, direct analysis requires much computation time, and thus the side edge coupled region is divided in N sections and expressed as a cascaded connection of N sub-circuits consisting of four-port asymmetrical coupled-lines as shown in Fig. 6.15. This analytical technique provides a shorter computation time while maintaining accuracy for practical design of filter devices.

Fig. 6.16 shows the calculated interstage coupling factor k of a linear tapered-line resonator-pair, obtained by the method described in the Appendix. Round dots in the figure, indicating measured data for a case of

$R_Z = 2$ and $Z_2 = 50$, show good agreement with calculated results, and thus proves the validity of the analysis method described. The proposed side edge coupling method provides a long coupling length, and consequently a strong coupling is obtained. This implies the availability of this resonator structure to realize BPF possessing wide pass-band [7].

Based on the above results, an experimental BPF was designed and fabricated. Design conditions are as follows:

Center frequency	:	$f_0 = 2.5\,\text{GHz}$
Pass-band bandwidth	:	$W > 200$ MHz
Pass-band VSWR	:	$V_{\text{SWR}} < 1.2$
Attenuation	:	$L_{\text{S}} > 20\,\text{dB}$ at $f_0 \pm 400\,\text{MHz}$.

Considering these specifications, a 3-stage BPF based on a Chebyshev response was chosen as filter design parameters. Thus, element values g_j are given as,

$$g_0 = g_4 = 1.0,$$
$$g_1 = g_3 = 0.779,$$
$$g_2 = 1.071.$$

As described in the Appendix, the coupling parameters of the filter are obtained from g_j as Q_{e} and $k = 0.086$. Electrical parameters of the resonator are given as $Z_1 = 25, Z_2 = 50, (R_Z = 2.0)$, while a substrate possessing a relative dielectric constant of $\varepsilon_{\text{r}} = 10.5$, and thickness H=1.27 mm was applied. Thus, the physical parameters of the resonator are obtained as,

$$W_1 = 3.74\,\text{mm}, \quad W_2 = 1.10\,\text{mm},$$
$$L_0 = 24.1\,\text{mm}.$$

Furthermore, the coupling circuit parameters are obtained from Fig. 6.13 and Fig. 6.16 as,

$$\ell_{\text{t}} = 4.2\,\text{mm},$$
$$S/H = 0.84(S = 1.1\,\text{mm}).$$

Finally, considering a slight shift in center frequency due to the coupling circuits, the resonator length was corrected through strict calculations based on electromagnetic analysis.

Figure 6.17 shows a photograph of the trial filter. Filter size measures 60 mm in length and 17 mm in width. Fig. 6.18 compares measured (solid line) and calculated (dotted line) frequency response near pass-band. Results prove close agreement, thus illustrating the validity of the design method. Pass-band insertion losses of less than 1.0 dB were obtained, and low-loss characteristics were achieved. Measured wide-band transmission characteristics are illustrated in Fig. 6.19. Spurious responses are generated at 4.4 and 6.64 GHz, both corresponding to expected values which were designed at frequency points shifted from the integer multiples of fundamental resonance

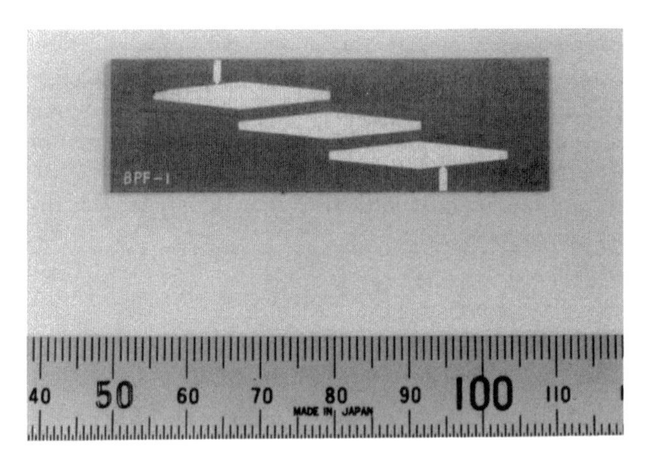

Fig. 6.17. Photograph of the experimental BPF using LTLRs

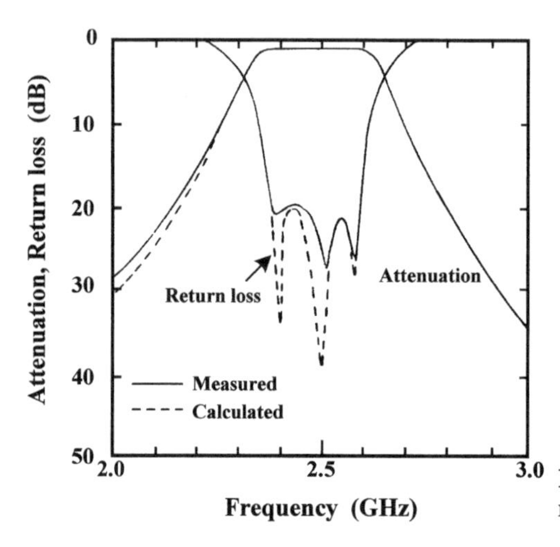

Fig. 6.18. Measured and calculated responses of the experimental BPF

frequency f_0. In addition, an attenuation of over 30 dB at $2f_0$ and $3f_0$ was obtained, these results illustrating the availability of this filter as a harmonics suppression BPF. These features resemble the previously described two-step SIR, and from these results we understand that the basic features of SIR are preserved in LTLR.

Although the above discussions are based on a stripline resonator structure possessing two open-ends as a linear tapered-line resonator (LTLR) example, a $\lambda_g/4$ type resonator with one short-circuited end can equally be applied in the RF bands, where a firm short-termination can be realized and generated parasitic components can be ignored. Moreover, this resonator structure becomes extremely practical when considering miniaturization. Figure 6.20 illustrates a filter example based on this resonator structure. It is

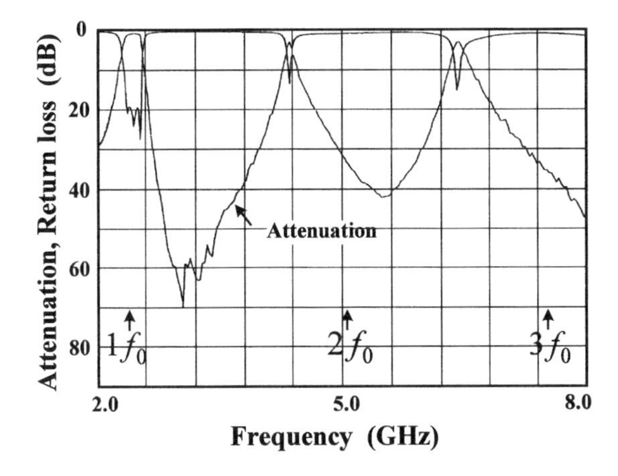

Fig. 6.19. Spurious responses of the experimental BPF

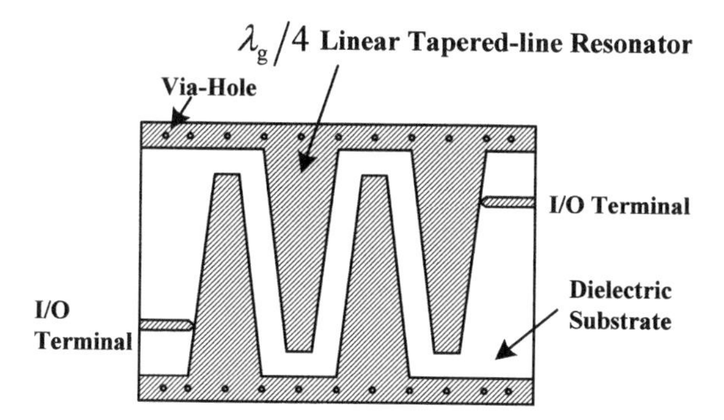

Fig. 6.20. Structure of an interdigital BPF using linear tapered-line resonators

obvious that the design method described above can be applied directly to filter design using a $\lambda_g/4$ type LTLR.

The hairpin resonator with internal coupling, discussed in Sect. 4.2 and shown in Fig. 6.21, obtains practical advantages of design flexibility and low-loss properties by introducing a resonator structure constructed of three types of transmission lines with different characteristic impedance. In addition, this resonator can be extended to a more generalized resonator structure as shown in Fig. 6.22, by applying a tapered-line structure to the internal coupled-lines and the single line. This structure also can be considered as an expanded concept of SIR.

6.3 Folded-Line SIR

The basic structure of SIR is composed of serial-connected plural transmission lines with different characteristic impedance. A single transmission line in a SIR structure can be replaced by a combination of transmission lines, allowing

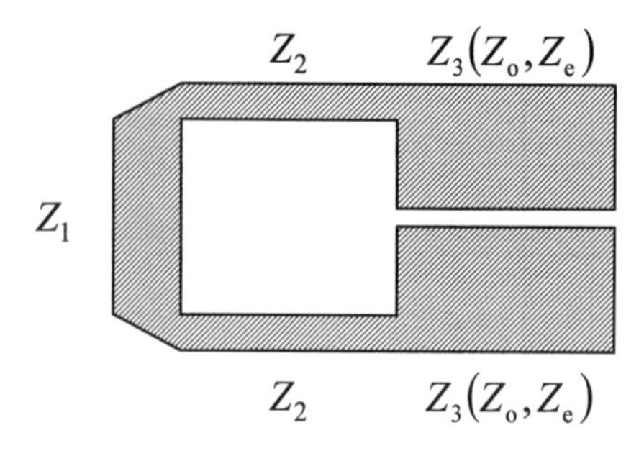

Fig. 6.21. Structure of a hairpin-type SIR composed of three transmission lines having different impedance

Fig. 6.22. Structure of a tapered-line resonator possessing inner coupled section

for new features while maintaining the advantages of a conventional SIR structure. An example of this resonator type is the folded-line resonator [8], which can be considered as one of the expanded concepts of the basic SIRs.

Figure 6.23a shows a conventional SIR structure. This SIR becomes equivalent to the structure illustrated in (b) when the transmission line including the open-end in (a) is divided into two transmission lines under the condition of $Z_2' = 2Z_2$ and an equal line length. In practical design, these two lines are folded as shown in (c) to enable miniaturization. The electrical performance of this resonator structure is equivalent to the conventional SIR shown in (a). However, in order to achieve further miniaturization, the coupling effect between the transmission lines Z_1 and Z_2' must be taken into consideration. Furthermore, to obtain an accurate resonance frequency estimation, effects of the T-junction and right-angle bends must be accurately calculated, thus requiring electromagnetic analysis based on methods such as the finite-element method.

Figure 6.24 shows a structural example of a three-stage BPF using folded-line SIR. Although this structure is not suitable for a wide-band application due to structural restrictions making it difficult to obtain a large I/O coupling, a narrow-band filter structure with a relative bandwidth of several per cent can be achieved, providing advantages of miniaturization.

While the above illustrates an example of folded-line SIR with two open-circuited ends, $\lambda_g/4$ type SIR with one open-circuited end also possess attractive applications as follows. One example is the RF short-circuited stub

(a) Basic SIR

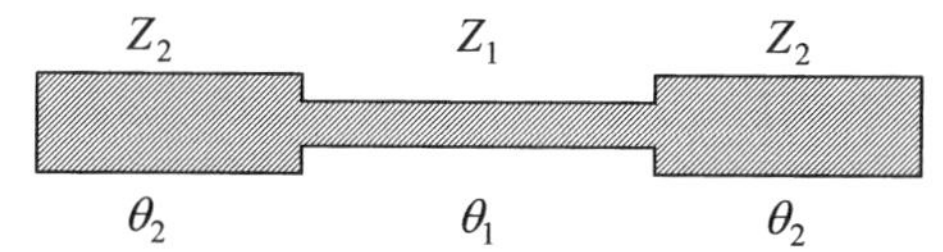

(b) An equivalent circuit of (a)

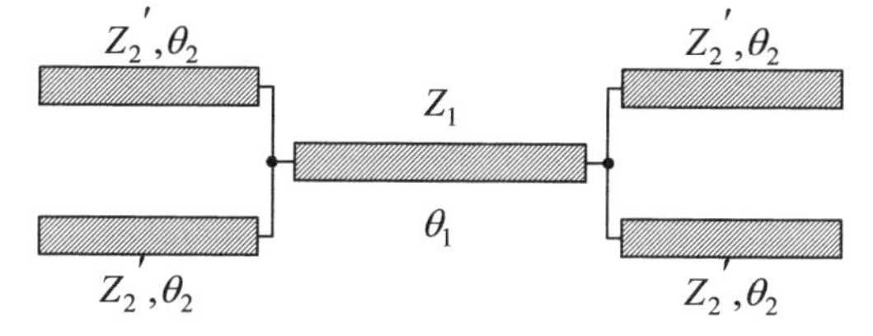

(c) Folded-line SIR

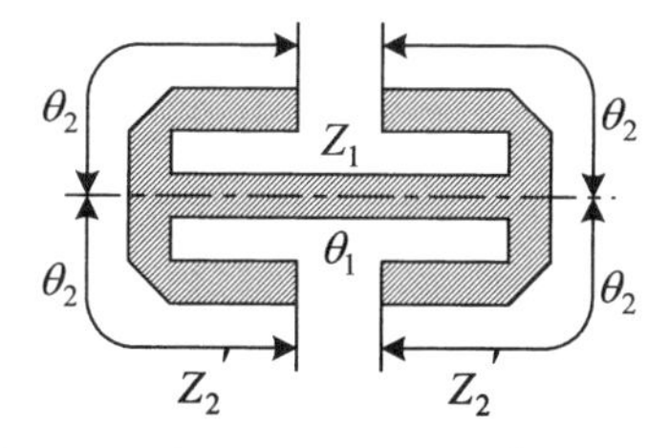

Fig. 6.23. Structural variations of $\lambda_g/2$ type SIRs. (**a**) Basic SIR. (**b**) An equivalent description of (**a**). (**c**) A folded-line SIR

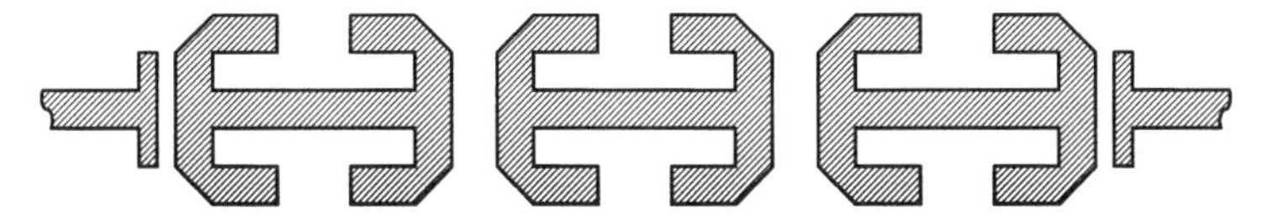

Fig. 6.24. BPF structure example using folded-line SIRs

shown in Fig. 6.25. Via-holes, basically applied for short circuits in the RF and microwave region, work well for a DC short circuit; however, circuit behavior in the RF band tends to become unstable due to stray inductance generated at the via-holes. To overcome this problem, short-circuit stubs, operating as short circuits at the desired RF frequency bands, are frequently employed. An open-ended $\lambda_g/4$ type transmission line or a radial line structure is often applied for such stubs, while both structures possess disadvantages, namely a narrow band for the former, and a large physical size for the latter. These drawbacks can be eased by applying a short-circuited stub using folded-line SIR as shown in Fig. 6.25, which possesses a wider band as compared to the $\lambda_g/4$ type stub while maintaining a far more compact physical structure as compared to the radial line stub. Assuming Z_{is} as the input impedance observed from the open-end of transmission line Z_1 in Fig. 6.25, Z_{is} is obtained

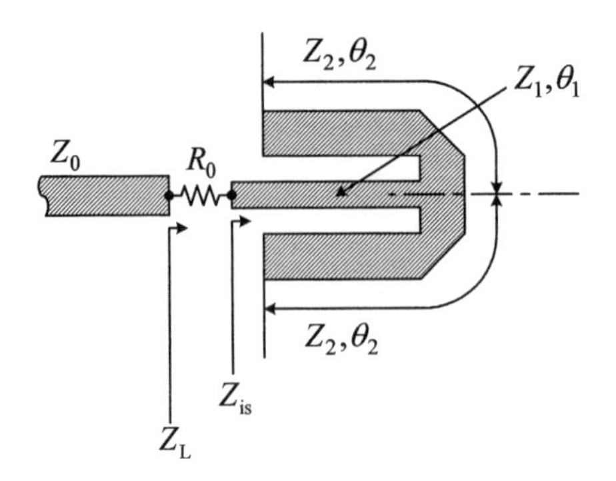

Fig. 6.25. RF short-circuit using folded-line SIR

as,

$$\begin{aligned} Z_{\text{is}} &= -\text{j}Z_1 \frac{2Z_2 - Z_1 \tan\theta_1 \cdot \tan\theta_2}{2Z_2 \tan\theta_1 + Z_1 \tan\theta_2} \\ &= -\text{j}Z_1 \frac{2R_Z - \tan\theta_1 \cdot \tan\theta_2}{2R_Z \tan\theta_1 + \tan\theta_2} \\ &\quad \text{where} \qquad R_Z = Z_2/Z_1. \end{aligned} \tag{6.3}$$

Z_{is} becomes zero under the condition:

$$2R_Z = \tan\theta_1 \tan\theta_2,$$

which is equivalent to the resonance condition of SIR, and thus a folded-line stub functions as a short-circuit near frequencies satisfying this condition. By connecting an ideal terminating resistor R_0 to the input port of the short-circuit stub, the input impedance seen from the other end of R_0, represented as Z_L, can be expressed as,

$$Z_L = R_0 + Z_{\text{is}}. \tag{6.4}$$

Let the normalized impedance $Z_0 = R_0 = 50\,\Omega$ and $\theta_1 = \theta_2$, the electrical angle θ_W allowing a standing wave ratio VSWR of less than 1.2 is obtained as follows.

$$\theta_L \leq \theta_W \leq \theta_H \tag{6.5}$$

$$\begin{aligned} \text{where} \quad \theta_L &= \tan^{-1}\left(\frac{A - B}{4Z_1}\right) \\ \theta_H &= \tan^{-1}\left(\frac{A + B}{4Z_1}\right) \\ A &= \sqrt{\left(\frac{50}{\sqrt{30}}\right)^2 (2 + R_Z)^2 + 8Z_1^2 R_Z} \\ B &= \frac{50}{\sqrt{30}}(2 + R_Z)^2. \end{aligned}$$

In addition, relative bandwidth is obtained as:

$$w = \frac{\theta_H - \theta_L}{\theta_0} \tag{6.6}$$

where $\theta_0 = \tan^{-1}\sqrt{R_Z/2}$.

Figure 6.26 illustrates actual calculated results. Results suggest that the bandwidth, which increases for a smaller R_Z and lower Z_1, is substantially wider than that of a $\lambda_g/4$ type uniform transmission line.

Utilizing the compact structure and wide-band properties of folded-line SIR, a LPF with excellent cut-off characteristics and a wide stop-band have been reported. Figure 6.27 shows an example of an experimental LPF employing six stubs based on two different folded-line SIR structures. The circuit behavior of these stubs changes from short-circuit to open-circuit in accordance with frequency, and stop-band width is expanded by changing the

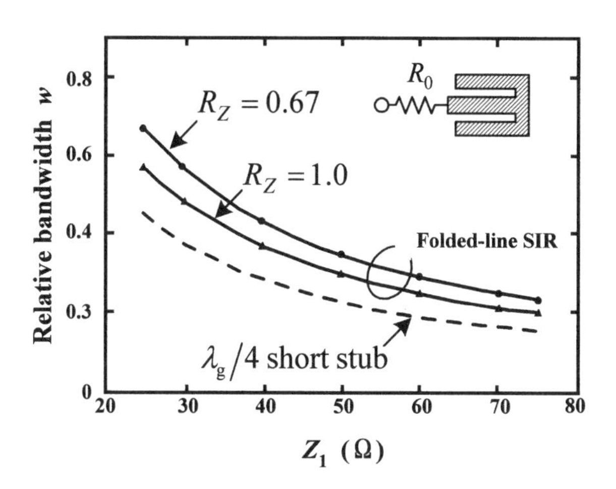

Fig. 6.26. Calculated results of relative bandwidth of 50 Ω termination using folded-line SIR

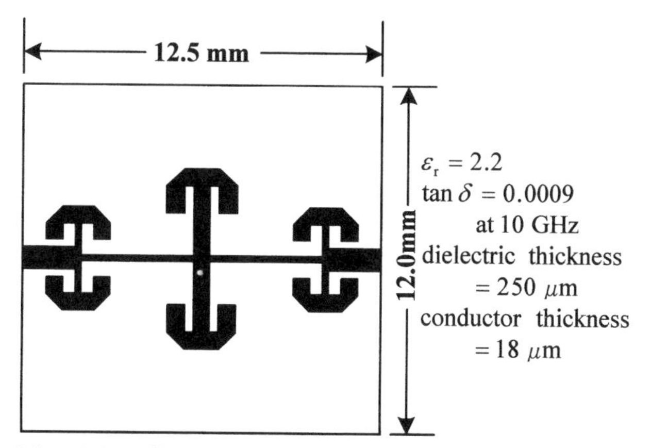

Fig. 6.27. Circuit pattern layout of the experimental LPF using folded-line SIRs

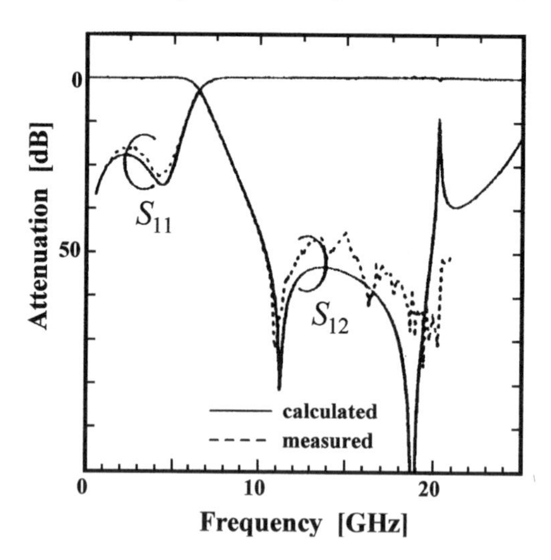

Fig. 6.28. Calculated and measured responses of the experimental LPF

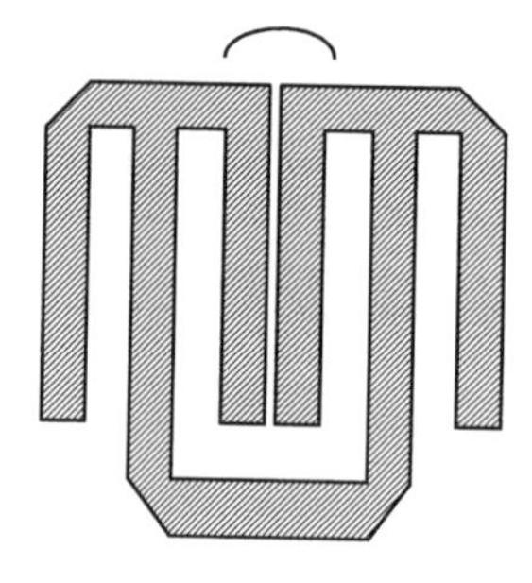

Fig. 6.29. Folded-line SIR with internal coupled-lines

frequencies satisfying the short-circuit condition between the central stubs and the stubs located at the ends of the BPF. Steep cut-off and wide stop-band characteristics are simultaneously obtained by applying a narrow-band design for the central stubs and a wide-band design for remaining.

Figure 6.28 indicates experimental results of the trial LPF. The pass-band is below 5 GHz, where an insertion loss of less than 0.5 dB and a return loss of 17 dB are obtained. Results show good agreement with designed values in the lower frequency bands, while at frequencies above 10 GHz, only an overall trend can be confirmed due to a decline in measurement accuracy for attenuation values close to 50 dB.

Although not mentioned in the above discussions, the basic concept of the folded-line resonator can equally be applied to a resonator structure possessing an internal coupling as described in Sect. 4.2. Figure 6.29 illustrates one such example. Although this resonator structure shows physical complexity, it is known to be suitable for miniaturization because the structure enables a reduction of impedance ratio. In addition, an extended design flexibility leaves hope for achieving a high Q value while maintaining a highly compact design.

6.4 Technological Trends of SIR in the Future

In the previous chapters we have discussed the various resonator structures that can be realized by expanding the basic concept of SIR. Although substantial progress has been achieved in systematic studies of the SIR, several problems remain yet to be solved. Our discussions in this section focus on four such topics.

Our first topic considers the circuit configuration of the SIR. Discussions up to this point treat the SIR as a serial connection of two or more transmission lines of different impedance, and analysis is basically focused on one-dimensional circuit theory. However, by applying a circuit configuration based on a composition of two-dimensional (planar) circuits and one-dimensional circuits, a desired resonator structure can equally be realized. One such example is a $\lambda_g/2$ type SIR accompanying two open-ends with an impedance ratio R_Z of less than 1. The width of the transmission line at an open-end inevitably increases in accordance with a reduction in impedance ratio R_Z, and thus a two-dimensional electromagnetic field distribution must be considered when analyzing the circuit behavior of such resonator structures. Expanding on this concept, we consider a resonator structure composed of one and two-dimensional circuits such as the examples shown in Fig. 6.30. While such resonator structures draw much interest, fundamental circuit behavior is yet to be clarified, and practical applications have not been discussed. Detailed studies on such resonator structures are expected in the future.

The next topic concerns the optimum structure of the SIR. Despite the numerous demands concerning resonator property, from a circuit design standpoint, miniaturization and high Q values are basically the most common and important requirements for a resonating circuit. Thus, determination of the optimal structure allowing a low resonance frequency and maximum Q value for a given (or more often limited) resonator size becomes an interesting and practical issue. For a coaxial type SIR, we derive Sect. 3.1 the conditions giving the maximum Q value based on an optimum relationship between the

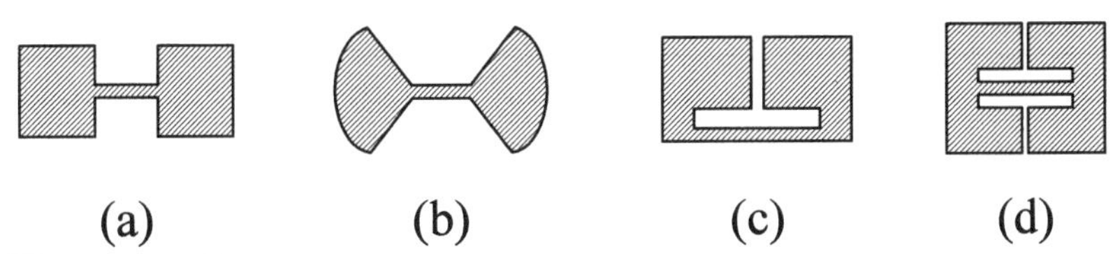

Fig. 6.30. Resonator structures composed of combination of one- and two-dimensional circuits

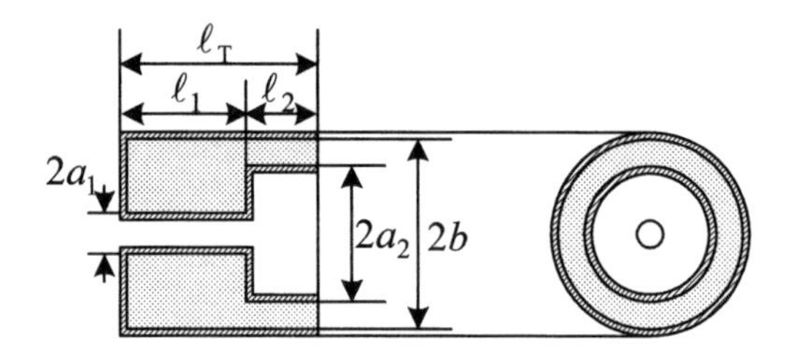

Resonator Parameters (condition: $\ell_T = 2b$)		Optimized	Conventional $(\ell_1 = \ell_2)$
Step-junction position	ℓ_1/ℓ_T	0.85	0.50
Radius ratio I	a_2/b	0.75	0.75
Radius ratio II	a_1/a_2	0.32	0.35
Norm. resonator length	ℓ_T/ℓ_0	0.59	0.54
Norm. unloaded Q	Q_{CN}	0.797	0.70
Figure of merit	Q_{CN}/θ_T	1.34	1.29

Fig. 6.31. Optimum structure of $\lambda_g/4$ type dielectric coaxial SIR

impedance of the two transmission lines. Conditions for step-junction position giving maximum Q values have also been reported.

Figure 6.31 shows an example of an optimized design for coaxial SIR obtained from electromagnetic field analysis based on a finite element method [9]. A lower resonance frequency and a higher unloaded-Q value are preferable when comparing electrical performance between resonators of identical size, and to conduct such assessments, we introduce an index value defined as the unloaded-Q value divided by the resonator electrical length θ_T (Q_0/θ_T). Optimized results in Fig. 6.31 are based on conditions of $2b = \ell_T$, while conventional SIR design, also shown in the figure, is based on conditions of minimum resonator length, namely $\ell_1 = \ell_2$. Results in the figure suggest that the index value can be improved from 1.29 to 1.34 by optimizing resonator structure. Although conditions $2b = \ell_T$ are applied for optimization of Q_C in this example, other conditions can equally be applied. However, few detailed and systematic studies on structural optimization of SIR have actually been conducted, and research on this issue is left for future assessment. In the case of a dielectric double coaxial resonator (DC-SIR) described in Sect. 3.4.4, the low impedance sections, namely ℓ_2 and ℓ_3, are restricted in length due to manufacturing conditions, and thus a structure possessing a relatively long ℓ_1 section is obtained. Although this structure is considered advantageous for high Q characteristics, strict analyses compared with conventional SIR are yet to be investigated. SIR structures applying planar circuits also pos-

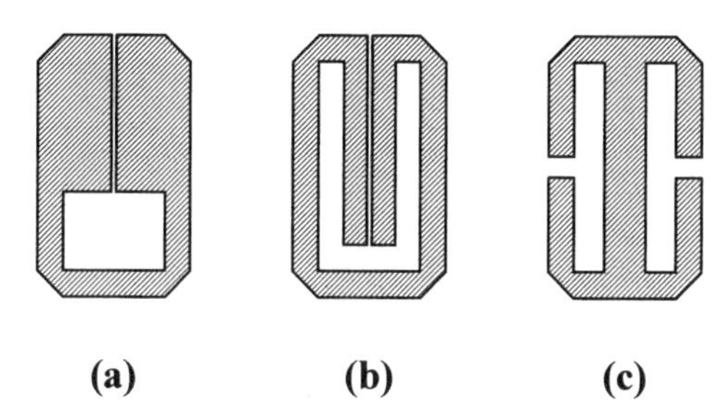

Fig. 6.32. Structural variations of planar SIR possessing the same outer-fringing shape

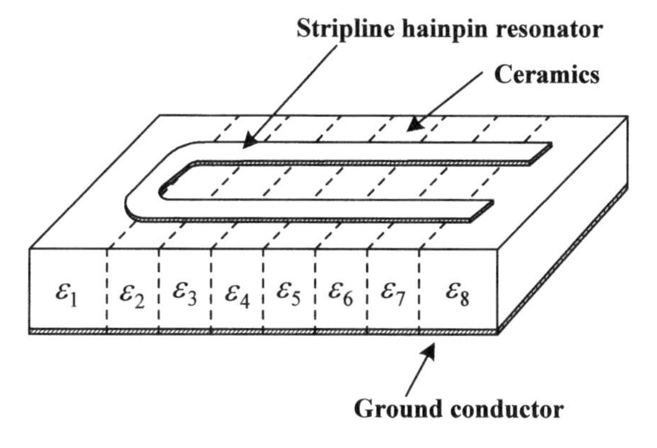

Fig. 6.33. Stripline SIR fabricated on a substrate with gradient dielectric constant

sess numerous structural variations as illustrated in Fig. 6.32, yet structural superiority concerning electrical performance has not been studied. It is an interesting technological topic to study the relationship between unloaded-Q and occupied circuit area indicating resonator size, comparing various types of SIR based on the same substrate and resonance frequency.

The next topic focuses on revolutionary SIR structures realized by the progress in present day material technology. Recent progress in ceramic processing technology has enabled substrate materials possessing gradient dielectric constant [10], as illustrated in Fig. 6.33. Reports reveal an experimental substrate possessing ten layers of different dielectric constant ranging from 17 to 60. Various SIR structures can be realized by introducing such functional materials, consequently improving features of compact size and high Q value. Two such examples, respectively based on a hairpin and coaxial resonator structure, are illustrated in Fig. 6.33 and Fig. 6.34. Although transmission-line width is constant for both structures, these resonators are classified as SIR due to a sequential change in transmission-line impedance. Further progress in processing technology allowing optional ceramic thickness

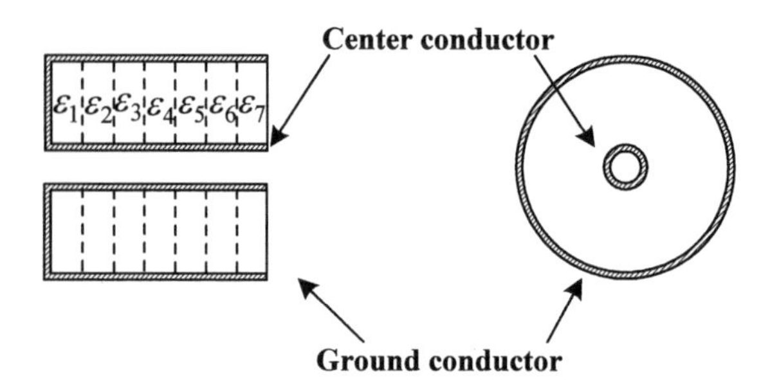

ε_i : Relative dielectric constant

$$\varepsilon_1 < \varepsilon_2 < \varepsilon_3 < \varepsilon_4 < \varepsilon_5 < \varepsilon_6 < \varepsilon_7$$

Fig. 6.34. Coaxial SIR using dielectric constant gradient material

of the individual layers is anticipated, leaving high expectations for various applications of this structure.

Our final topic concerning SIR future trends is the expansion of applicable frequency range. SIR R&D up to now have progressed mainly in frequency bands ranging from UHF (0.3–3 GHz) to SHF (3–30 GHz), and various products have proved their availability in these frequency bands. Although SIR application to the VHF band (0.03–0.3 GHz) is technically possible, when considering the availability of compact and high Q resonators such as SAW devices already developed in this frequency region, application is restricted to special scenes where SIR structure is inevitable. One such promising application is the high power filter. Although a dielectric resonator BPF could equally be applied for this purpose, the volume of the dielectric material required for this frequency band increases to a point where ceramic processing become extremely difficult. Under such processing restrictions, an air cavity type DC-SIR emerges as a powerful candidate, and when considering its metal-based structure allowing effective countermeasures against high-power handling accompanying heat, the structure seems to be a perfect solution for this restricted application.

In the millimeter-wave region (the EHF band: 30–300 GHz), applications based on one-wavelength ring-type SIR with low radiation loss and half-wavelength type SIR with hairpin or loop-shaped structure are expected to progress, as previously describeds in Chap. 5. Such developments require the proposal of various transmission-line structures based on micromachining technology. Furthermore, in addition to filtering devices, SIR application to resonance elements coexisting with active circuits within a MIC or MMIC are foreseen. In all respects, the SIR possesses an extremely wide application range as a transmission-line resonator structure, leaving high expectations for expansion of applicable frequency range and further research and development in the future.

Appendix. Analysis of Resonator Properties Using General-Purpose Microwave Simulator

A.1 Design Parameters of Direct-Coupled Resonator BPF

As previously described, bandpass filters in the RF and microwave regions are generally realized by the cascaded connection of plural resonators with uniform structure using appropriate coupling circuits. Starting with given filter specifications (e.g., center frequency, pass-band bandwidth, VSWR, insertion loss, attenuation at specified frequency, etc.), the first step in a filter synthesis procedure is to obtain the filter design parameters, such as the required stage number n, element values g_j, and unloaded-Q, after choosing the filter response type of Chebyschev or Butterworth. In accordance with these fundamental parameters, the next step is to calculate the electrical parameters and/or structural parameters based on the applied resonator structure.

Filter design based on lumped-element LC resonators and lumped-element coupling circuits allow direct calculation of the electrical parameters. Contrarily, in the high frequency regions such as RF and microwave bands, distributed-element circuits are often applied to filtering devices, and thus both electrical and structural parameters are required for filter design. Since filter design theories have been generalized and systematized based on lumped-element circuit theory, the electrical parameters are first calculated by approximating distributed circuits into lumped-element circuits, which are in turn converted into the structural parameters determining resonator design.

Filter design can easily be obtained when the conversion between electrical and structural parameters is analytically derived and expressed in the form of a mathematical equation. Yet for complicated geometrical structures, generalized conversion equations become difficult to derive, and thus experimental results are often combined with analytical expressions in order to obtain the structural parameters. Design methods in which filter design parameters are determined mainly by experimental results have also been established [1].

Recent CAD technology for microwave circuits has shown remarkable progress, allowing for accurate and highly reliable analysis of electromagnetic circuits possessing three-dimensional structures. While it is possible to develop an application-specific simulator for each individual filter structure, here we discuss a generalized method of obtaining design charts [2] required for filter synthesis by calculating the electrical performances of fundamental filter structures using commercially available general-purpose microwave simulators. The intentions of this approach are to determine filter parameters by replacing conventional experiment-based design processes with virtual experiments based on simulated results.

Circuits parameters required for the experimental method are; resonance frequency f_0, resonator unloaded-Q (Q_0), external-Q (Q_e) related to input/output couplings, and interstage coupling coefficient k which prescribes coupling between the resonators. The theoretical background and actual procedures of this method, where design parameters are determined by simulating circuit responses using a single resonator and a resonator-pair, are described as follows.

A.2 Filter Design by Experimental Method

Based on the initial design specifications of the BPF, center frequency f_0, number of stages n, relative bandwidth w, element values g_j, and insertion loss L_0(dB) are assumed to be given. Resonator structure, resonance frequency (center frequency), unloaded-Q(Q_0), and external-Q(Q_e) can be obtained by examining the single resonator. These parameter can also be experimentally measured.

As discussed in Sect. 3.2.1, insertion loss is determined by Q_0, which must be estimated under the following conditions.

$$Q_0 > \frac{4.434}{w \cdot L_0} \sum_{j=1}^{n} g_j. \tag{A.1}$$

The unloaded-Q (Q_0) is determined by the geometrical structure of the distributed-element circuit, and thus the relationship between resonator structure and Q_0 must be thoroughly investigated because it becomes an extremely important parameter determining the total size of the BPF.

The external-Q (Q_e) is an important parameter which provides the relationship between the filter and the external circuit (input and output circuit), thus providing impedance matching between them. They are expressed as [1],

$$Q_{e1} = \frac{g_0 g_1}{w} \tag{A.2}$$

$$Q_{e2} = \frac{g_n g_{n+1}}{w} \tag{A.3}$$

Q_{e1} and Q_{e2} indicate the external-Q at the input and output port, respectively. When the input and output impedance are equal, then $Q_{e1} = Q_{e2}$.

The interstage coupling coefficient k has the following relationship with the basic design parameters.

$$k_{j,j+1} = \frac{w}{\sqrt{g_j g_{j+1}}} \tag{A.4}$$

$k_{j,j+1}$indicates the coupling factor between the j-th and (j+1)-th resonator. The k value can easily be determined by obtaining the frequency response of a resonator-pair.

From the above discussions, we understand that filter design can be achieved by considering experimentally measurable values of f_0, Q_0, Q_e, and k. Consequently, it is possible to synthesize the BPF by obtaining the relationship between these parameters and the physical structure of the filter.

A.3 Determination of Q and k Using General Purpose Microwave Simulator

A.3.1 Determination of Q

The quality factor of a resonator is expressed as three Q values of different definition: the unloaded-Q (Q_0) which indicates the figure of merit of the resonator itself; the external-Q (Q_e) which expresses the coupling condition between the external circuits and the resonator; and the loaded-Q (Q_L) which represents the total Q including resonator and external circuits. The following relationship is recognized between these values.

$$\frac{1}{Q_L} = \frac{1}{Q_e} + \frac{1}{Q_0} \tag{A.5}$$

This relationship implies that by obtaining two of the three Q values, the remaining can be calculated from (A.5).

Experimental methods for Q measurement have been studied by Ginzton [3], and these results have laid the foundation for the reflection method or one-port method. In this section an alternative method, namely the transmission method or two-port method, is employed. This method is suitable for computer simulations where two-port symmetrical circuits are utilized to simultaneously obtain Q_0 and Q_e values, while for experimental methods obtaining these values accurate results cannot be expected due to increased error factors.

Figure A.1 shows a generalized two-port symmetrical circuit for the single resonator. Measurements obtained from actual experimental circuits often lack accuracy due to processing limits which make the symmetrical circuit difficult to fabricate. Such problems can be ignored when conducting virtual experiments based on circuit simulators. In Fig. A.1, the resonator itself is indicated as $Y(\omega)$, while the coupling circuits are expressed by the admittance inverter J_0, which indicates a generalized coupling parameter. For practical simulations, the resonators and coupling circuits must be described as actual circuit element values or physical parameters in a descriptive language understood by the applied simulator. As in an experimental method where Q_0

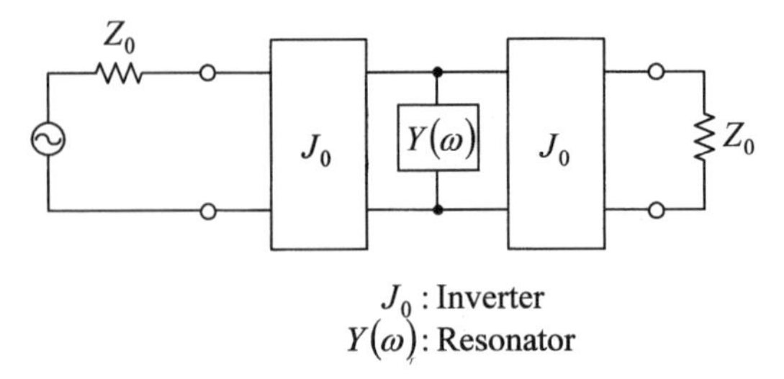

Fig. A.1. Circuit configuration for determining f_0, Q_0 and Q_e

and Q_e are obtained by measuring frequency response, for this method the transmission response of the circuit shown in Fig. A.1 must be calculated. For simplicity, $Y(\omega)$ is expressed as follows.

$$\begin{aligned} Y(\omega) &= G_0 + jB(\omega) \\ &= G_0 + jb_0(\omega/\omega_0 - \omega_0/\omega) \end{aligned}$$

where

G_0 : conductance of the resonator,
b_0 : slope parameter of the resonator,
ω_0 : angular resonance frequency.

This expression illustrates a lumped-element approximation of the resonator at frequencies near resonance. Furthermore, a G_0 value independent of frequency is assumed. Considering analysis within a sufficiently narrow frequency band, the admittance inverter J_0 is also assumed to be independent of frequency and is thus treated as constant, although strictly it will possess frequency-dependent characteristics. Under these conditions, we first obtain the total F-matrix $[F_t]$ of the circuit shown in Fig. A.1, and next derive transducer loss $L(\omega)$, defined as the ratio of power available from the source and power absorbed by the load.

$$\begin{aligned} [F_t] &= \begin{bmatrix} 0 & j/J_0 \\ jJ_0 & 0 \end{bmatrix} \begin{bmatrix} 1 & 0 \\ Y(\omega) & 1 \end{bmatrix} \begin{bmatrix} 0 & j/J_0 \\ jJ_0 & 0 \end{bmatrix} \\ &= \begin{bmatrix} -1 & -G_0/J_0^2 - jb_0(\omega/\omega_0 - \omega_0/\omega)/J_0^2 \\ 1 & -1 \end{bmatrix} \\ &= \begin{bmatrix} A_t & B_t \\ C_t & D_t \end{bmatrix}. \end{aligned}$$

Using the above matrix elements,

$$\begin{aligned} L(\omega) &= \frac{1}{4Z_0^2} \left| A_t Z_0 + B_t + C_t Z_0^2 + D_t Z_0 \right|^2 \\ &= \frac{1}{4Z_0^2} \left\{ (2Z_0 + G_0/J_0^2)^2 + b_0^2(\omega/\omega_0 - \omega_0/\omega)^2/J_0^4 \right\}. \end{aligned} \tag{A.6}$$

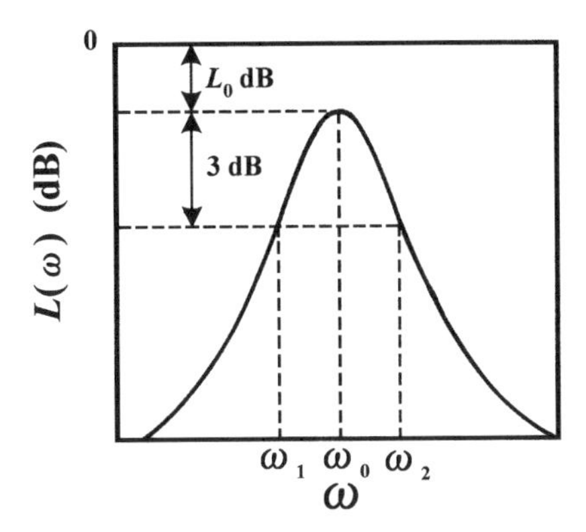

Fig. A.2. Frequency response of a single resonator test circuit

Considering $Q_0 = b_0/G_0$ and $Q_e = b_0/Z_0 J_0^2$,

$$L(\omega) = \frac{1}{4}\left(2 + \frac{Q_e}{Q_0}\right)^2 + Q_e^2\left(\frac{o}{\omega_0} - \frac{\omega_0}{\omega}\right)^2 . \tag{A.7}$$

Thus, as illustrated in Fig. A.2, the frequency response of $L(\omega)$ will feature a single peak characteristic. $L(\omega)$ attains a maximum value L_0 at ω_0, which can be obtained from (A.7) by substituting $\omega = \omega_0$ as,

$$L_0 = 1 + \frac{Q_e}{Q_0} + \frac{1}{4}\left(\frac{Q_e}{Q_0}\right)^2 . \tag{A.8}$$

ω_1 and $\omega_2 (\omega_1 < \omega_2)$ are defined as the angular frequencies showing a 3 dB downfall from L_0. Since ω_1 and ω_2 exist near the resonance angular frequency ω_0, $\omega_0 - \omega_1 = \omega_2 - \omega_0$ is assumed, and thus by introducing Ω_1 as,

$$\Omega_1 = \left|\frac{\omega_1}{\omega_0} - \frac{\omega_0}{\omega_1}\right|$$

$L(\omega_1)$ is expressed as,

$$\begin{aligned} L(\omega_1) &= 2L_0 \\ &= 1 + \frac{Q_e}{Q_0} + \frac{1}{4}\left(\frac{Q_e}{Q_0}\right) + \frac{1}{4}Q_e^2 \Omega_1^2 . \end{aligned} \tag{A.9}$$

Considering (A.8) and (A.9),

$$L_0 = \frac{1}{4} Q_e^2 \cdot \Omega_1^2 .$$

Consequently, we obtain,

$$Q_e = \frac{2}{\Omega_1}\sqrt{L_0} \tag{A.10}$$

$$Q_0 = \frac{\sqrt{L_0}}{\Omega_1\left(\sqrt{L_0} - 1\right)} . \tag{A.11}$$

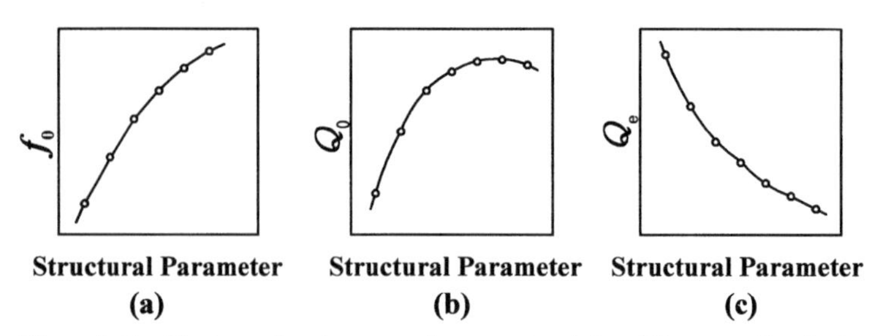

Fig. A.3. Design chart examples required for filter synthesis by experimental method. (**a**) Resonance frequency. (**b**) Unloaded-Q. (**c**) External-Q

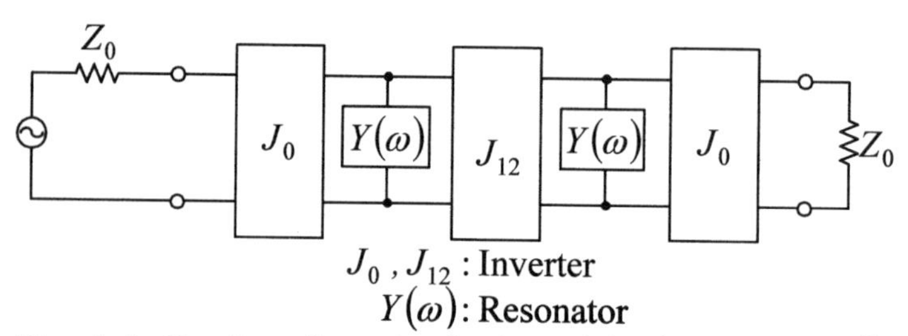

Fig. A.4. Circuit configuration for determining interstage coupling factor

The above discussions suggest that the required information on f_0, Q_0 and Q_e can be obtained from the calculated frequency response of the circuit shown in Fig. A.2 using a microwave circuit simulator. Preparations of the design charts exemplified in Fig. A.3 prove highly efficient time-use for actual filter synthesis procedures.

A.3.2 Determination of Coupling Coefficient

Resonator-pair measurements are a conventional method to experimentally obtain resonator coupling coefficients [1], where the two peaks observed in the transmission response of a circuit consisting of two identical resonators are measured. We start our discussions from a resonator-pair based on a structure applied to the basic configuration of the BPF, and by expressing this circuit with its inverter parameters, we derive a method for determining coupling factor [2] suitable for circuit simulations. Figure A.4 illustrates the circuit configuration of the target resonator-pair. J_0 and J_{12} represent the inverter parameters for I/O and interstage coupling, while the resonator is expressed as $Y(\omega)$. As discussed in Sect. A.3.1 total F-matrix and transducer loss $L(\omega)$ are derived as follows,

$$
\begin{aligned}
[F_\mathrm{t}] &= \begin{bmatrix} A_\mathrm{P} & B_\mathrm{P} \\ C_\mathrm{P} & D_\mathrm{P} \end{bmatrix} \\
&= \begin{bmatrix} -\mathrm{j}Y(\omega)/J_{12} & -\mathrm{j}J_{12}/J_0^2 - \mathrm{j}Y(\omega)^2/J_0^2 J_{12} \\ \mathrm{j}J_0^2/J_{12} & \mathrm{j}Y(\omega)/J_{12} \end{bmatrix} .
\end{aligned}
$$

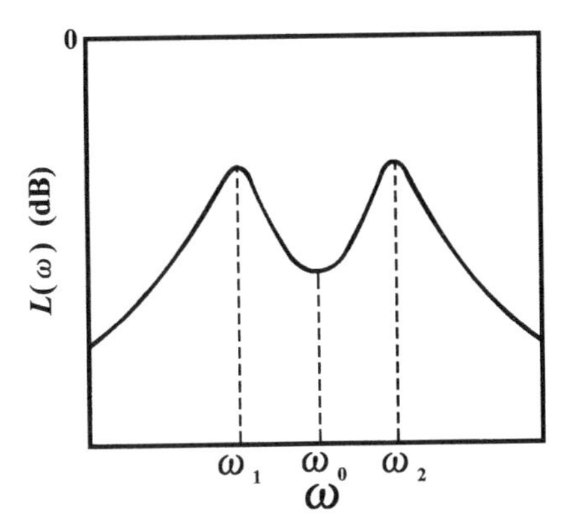

Fig. A.5. Frequency response of a resonator-pair

And thus,

$$L(\omega) = \frac{|A_P Z_0 + B_P + C_P Z_0 + D_P Z_0|^2}{4Z_0^2}$$
$$= \left[B(\omega)^4 - 2\left\{J_{12}^2 - (G_0 + J_0^2 Z_0)^2\right\}B(\omega)^2 + \left\{J_{12}^2 + (G_0 + J_0^2 Z_0)^2\right\}^2\right] \Big/ 4Z_0^2 J_0^2 J_{12}^2 \quad \text{(A.12)}$$

When $J_{12}^2 - (G_0 + J_0^2 Z_0)^2 > 0$, (A.12) has a frequency response possessing two peaks as shown in Fig. A.5. Although $J_{12} > G_0 + J_0^2 Z_0$ is not always achieved when conducting actual experiments, this condition can easily be satisfied for virtual experiments based on simulations, where J_0 can be reduced to any given value. A reduction in J_0 is equivalent to reducing I/O coupling strength, namely, to adopt a loose coupling. Let the angular frequencies of the two peaks be ω_1 and ω_2 $(\omega_2 > \omega_0 > \omega_1)$, and the following relationship is derived from the conditions giving peak values.

$$B(\omega_2) = -B(\omega_1) = \sqrt{J_{12}{}^2 - (G_0 + J_0^2 Z_0)^2} \quad \text{(A.13)}$$

Then,

$$\sqrt{J_{12}{}^2 - (G_0 + J_0^2 Z_0)^2} = \frac{1}{2}\left\{B(\omega_2) - B(\omega_1)\right\}.$$

Thus, J_{12} is expressed as,

$$J_{12} = \sqrt{\left\{B(\omega_2) - B(\omega_1)\right\}^2/4 + (G_0 + J_0^2 Z_0)^2}. \quad \text{(A.14)}$$

In addition,

$$B(\omega) = b_0\left(\frac{\omega}{\omega_0} - \frac{\omega_0}{\omega}\right)$$

b_0 : slope parameter of resonator.

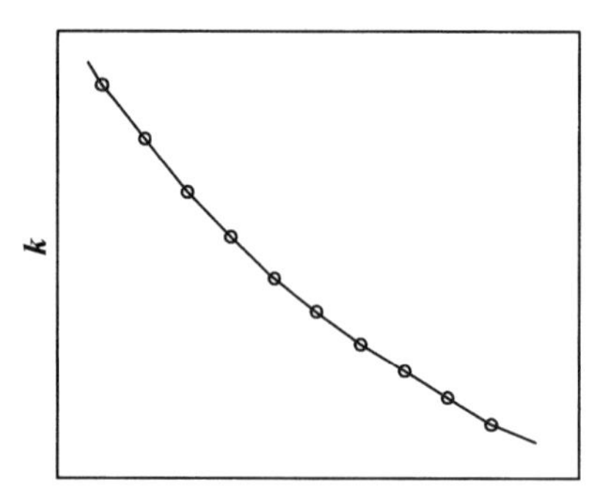

Fig. A.6. Design chart example required for determining interstage coupling

Thus,

$$B(\omega_2) - B(\omega_1) \cong 2b_0 \cdot \frac{\omega_2 - \omega_1}{\omega_0}.$$

Considering $J_{12} = b_0 k$, $G_0 = b_0/Q_0$, and $Q_e = b_0/Z_0 J_0^2$, (A.14) can be expressed as,

$$k \cong \sqrt{\left(\frac{\omega_2 - \omega_1}{\omega_0} +\right)^2 + \left(\frac{1}{Q_0} + \frac{1}{Q_e}\right)^2}. \tag{A.15}$$

Assuming $Q_0 >> 1$ and $Q_e >> 1$, we obtain,

$$k \cong \frac{\omega_2 - \omega_1}{\omega_0} = \frac{f_2 - f_1}{f_0}. \tag{A.16}$$

From the above discussions, it is possible to determine the coupling coefficient between resonators by obtaining the circuit response of a resonator-pair conditioned to generate two peaks. As described in Sect. A.3.1, for actual filter synthesis, preparation of design charts, which provide the relationship between the geometrical arrangement of the resonators and the coupling parameters as illustrated in Fig. A.6, are highly recommended from a practical point of view.

References

1. Introduction

1. G.L. Matthaei, L. Young, E.M.T. Jones : in *Microwave Filters, Impedance-Matching Networks, and Coupling Structures* (McGraw-Hill, New York, 1964) (Reprint version: Artech House, Inc., Dedham, Massachusetts (1980))
2. R.D. Richtmeyer : J. Appl. Phys., **10**, 391 (1939)
3. S.B. Cohn : IEEE Trans. MTT, **16**, 218 (1968)
4. G.L. Matthaei, G.L. Hey-Shipton : IEEE Trans. MTT, **42**, 1287 (1994)
5. W.P. Mason : Bell Sytem Tech. J., **13**, 405 (1934)
6. H. Matthews : in *Surface Wave Filters* (Wiley-Interscience Publication, New York, 1977)
7. J.F. Dillon : Phys. Rev., **105**, 759 (1957)
8. R.W. Hamming : in *Digital Filters* (Prentice-Hall, New York, 1977)
9. V.H. MacDonald : Bell Sytem Tech. J., **58,** 15 (1979)
10. L. Young(ed.) : in *Microwave Filters Using Parallel Coupled Lines* (Artech House Inc., Dedham, Massachusetts, 1972)
11. M. Makimoto, S. Yamashita : Proc IEEE, **67**, 16 (1979)

2. Basic Structure and Characteristics of SIR

1. A. Gopinath, A.F.Thomson, I.M.Stephenson: IEEE Trans. MTT, **24**, 142 (1976)
2. M. Makimoto, S.Yamashita : Proc. IEEE, **67**,16 (1979)
3. M. Makimoto, H.Endoh, S. Yamashita, K. Tanaka : in *1984 IEEE VTC Digest,* 201 (1984)
4. M. Sagawa, M. Makimoto, S.Yamashita : IEEE Trans. MTT, **45**, 1078 (1997)

3. Quarter-Wavelength Type SIR

1. M. Sagawa, M. Makimoto, S. Yamashita : IEEE Trans. MTT, **45**, 1078 (1997)

2. H.E. Green : IEEE Trans. MTT, **13**(5), 676(1965)
3. N. Marcuvitz : in *Waveguide Handbook* (McGraw-Hill, New York, 1951)
4. W.J. Getsinger : IRE Trans. MTT, **10**, 65 (1962)
5. J.R. Whinnery, H. W.Jamiesen : Proc. IRE, **32**, 98 (1944)
6. S.B. Cohn : Proc. IRE, **45**, 187 (1957)
7. G.L. Matthaei, L. Young, E.M.T. Jones : in *Microwave Filters, Impedance-Matching Networks, and Coupling Structures* (McGraw-Hill, New York, 1964)
8. M. Makimoto, H. Endoh, S. Yamashita, K. Tanaka : in *Proc. 1984 IEEE VTC,* 201 (1984)
9. M. Sagawa, M. Matsuo, M. Makimoto, K. Eguchi : IEICE Trans. Electron., **E78**-C, 1051 (1995)
10. R.M. Kurzrok : IEEE Trans. MTT, **14**, 351 (1966)
11. S. Kawashima, M. Nishida, I. Ueda, H. Ouchi : J. Am. Ceram. Soc., **66**(6), 421 (1983)
12. H. Ouchi, S. Kawashima : Jpn. J. Appl. Phys., **24**, Supplement 24-2, 60 (1985)
13. M. Sagawa, M. Makimoto, S. Yamashita : IEEE Trans. MTT, **33**, 152 (1985)
14. M. Sagawa, M. Makimoto, K. Eguchi, F. Fukushima : IEICE Trans. Electron., **E74**, 1221 (1991)
15. H. Matsumoto, T. Tsujiguchi, T. Nishikawa : in *1995 IEEE MTT-S Digest,* 1593 (1995)
16. H.C. Chung, C.C. Yeh, W.C. Ku, K.C. Tao : in *1996 IEEE MTT-S Digest,* 619 (1996)
17. I. Ishizaki, T. Uwano : in *1994 IEEE MTT-S Digest,* 617 (1994)

4. Half-Wavelength Type SIR

1. M. Makimoto, S. Yamashita : IEEE Trans. MTT, **28**, 1413 (1980)
2. G.L. Matthaei, L. Young, E.M.T. Jones : in *Microwave Filters, Impedance-Matching Networks, and Coupling Structures* (McGraw-Hill, New York, 1964)
3. S.B. Cohn : IRE Trans. MTT, **6**, 223 (1958)
4. H.M. Altschuler, A.A. Oliner : IRE Trans. MTT, **8**, 328 (1960)
5. V. Nalbandian, W. Steenarrt : IEEE Trans. MTT, **20**, 553 (1972)
6. E.G. Cristal, S. Frankel : IEEE Trans. MTT, **20**, 719 (1972)
7. A. Sheta, J.P. Coupez, G. Tanne', S. Toutain, J.P. Blot : in *1996 IEEE MTT-S Digest,* 719 (1996)
8. M. Sagawa, M. Makimoto, S. Yamashita : IEEE Trans. MTT, **33**, 152 (1983)
9. C.H. Ho, J.H. Weidman : Microwave System News (MSN), Oct. 88 (1983)
10. M. Makimoto, M. Sagawa : in *1986 IEEE MTT-S Digest,* 411 (1986)

11. S. Sagawa, I. Ishigaki, M. Makimoto, T. Naruse : in *1988 IEEE MTT-S Digest,* 605 (1988)
12. M. Sagawa, K. Takahashi, M. Makimoto : IEEE Trans. MTT, **42**, 1287(1989)
13. J.C. Rodult, A. Skrivervik, J.F. Zurcher : Microw. Opti. Tech. Lett., **4**, 384 (1991)
14. G.L. Matthaei, G.L. Hey-Shinpton : IEEE Trans. MTT, **42**, 1287 (1994)
15. G.L. Matthaei, N.O. Fenzi, R.J. Forse, S.M. Rohlfing : IEEE Trans. MTT, **45**, 226 (1997)
16. A. Enokihara, K. Setsune, K. Wasa, M. Sagawa, M. Makimoto : Electron. Lett., **28**, 1925 (1992)
17. A. Enokihara, K. Setsune : J. Supercond., **10**, 49 (1997)
18. A. Lander, D.W. Face, W.L. Holstein, D.J. Kountz, C. Wilker : in *Proc. 5th Int. Symp. on Superconductivity,* 925 (1992)
19. M. Sagawa, M. Makimoto, S. Yamashita : IEEE Trans. MTT, **45**, 1078 (1997)
20. H. Yabuki, M. Sagawa, M. Makimoto : IEICE Trans. Electron., **E76-C**, 932 (1993)

5. One-Wavelength Type SIR

1. I. Wolff : Electron. Lett., **8**, 163 (1972)
2. M. Guglielmi, G. Gatti : in *Proc. 1990 European Microwave Conf.,* 901 (1990)
3. U. Karacaoglu, I.D. Robertson, M. Guglielmi : in *Proc. 1994 Eupropean Microwave Conf.,* 472 (1994)
4. U. Karacaoglu, D. Sanchez-Hernandez, I.D. Robertson, M.Guglielmi : in *1996 IEEE MTT-S Digest,* 1635 (1996)
5. H. Yabuki, M. Sagawa, M. Matsuo, M. Makimoto : IEEE Trans. MTT, **44**, 723 (1996)
6. M. Sagawa, M. Makimoto, S. Yamashita : IEEE Trans. MTT, **45**, 1078 (1997)
7. K. Takahashi, S. Fujita, U. Sangawa, A. Ono, T. Urabe, S. Takeyama, H. Ogura, H. Yabuki : in *1999 IEEE MTT-S Digest,* 229(1999)

6. Expanded Concept and Technological Trends of SIR

1. M. Sagawa, M. Makimoto, S. Yamashita : IEEE Trans. MTT, **45**, 1078 (1997)
2. S.K. Lim, H.Y. Lee, J.C. Kim, C. An : IEEE Microw. Guid. Wave Lett., **9**, 143 (1999)
3. M. Makimoto, S. Yamashita : Proc. IEEE, **67**, 16 (1979)

4. S. Yamashita, M. Makimoto : IEEE Trans. MTT, **31**, 485 (1983)
5. S. Yamashita, M. Makimoto : IEEE Trans. MTT, **31**, 679 (1983)
6. S. Yamashita, K. Tanaka, H. Mishima : Proc. IEEE, **67**, 1666 (1979)
7. M. Sagawa, H. Shirai, M. Makimoto : IEICE Trans. Elec., **E76-C**, 985 (1993)
8. Y. Qian, K. Yanagi, E. Yamashita : in *Proc. 1995 Eropean Micowave Conf.*, 1209 (1995)
9. M. Toki, H. Arai, M. Makimoto : IEICE Trans. C-1, **J75-C-1**, 503(1992)
10. S. Kamba, K. Kawabata, H. Yamada, H. Takagi : Denshi Tokyo (IEEE Tokyo Section), **No.33**, 49 (1994)

Appendix

1. G.L. Matthaei, L.Young, E.M.T.Jones : in *Microwave Filters, Impedance Matching Networks, and Coupling Structures* (McGraw-Hill, New York, 1964)
2. M.Sagawa, K.Takahashi, M.Makimoto : IEEE Trans. MTT, **37**, 1991 (1989)
3. E.L.Ginzton : in *Microwave Measurements* (McGraw-Hill, New York, 1957)

Index

active filters 2
admittance inverter 69
air-cavity type resonator 39
antenna duplexer 43
antiresonance 101
asymmetrical coupled-lines 136
attenuation poles 62, 116

balanced mixer 103
BEF 3, 42
BPF 3
bulk-wave filters 2

ceramic materials 45
ceramic resonator 2
characteristic impedance 7, 12
coaxial resonators 6
conductivity 26
coupled-lines 73, 74
coupling coefficients 154
crystal resonator 2

DC-SIR 39, 53
dielectric coaxial resonators 47, 53
dielectric DC-SIR 53
dielectric filter 2
dielectric materials 45, 128
dielectric monoblock filters 58
dielectric resonator 1, 6, 45
digital filters 3
diplexer 112
discontinuity 21, 113, 114
distributed-element 16, 30
dual-mode filter 113, 119
duplexer 3, 42

effective dielectric constant 124
element value g_j 30, 33
equivalent circuit 18, 39, 69, 90, 115
even-mode excitation 115
external-Q (Q_e) 91, 135, 150

filter specifications 149
filter synthesis 149
folded-line SIR 140
fringing capacitance 21
fundamental matrix 68

gap capacitor 35
gradient dielectric constant 147

hairpin-shaped SIR 100
half-wavelength resonator 7
helical resonators 5
high power filter 148
HPF 3
HTS filter 98

impedance matrix 68
impedance ratio 11, 13, 21
inhomogeneous impedance 127
insertion loss 33, 99
interdigital capacitor 35
interference signals 4
internally coupled SIR 84
inverted micro-stripline 122
inverter parameter 31, 35

LC resonators 1, 149
line length correction 32
linear tapered-line resonator 131
loaded-Q 151
loss tangent 24
LPF 3, 129, 143
lumped-element 16, 30

magnetostatic mode filters 5
micro-machining technology 122, 148
micro-stripline 7, 60
millimeter wave IC 122

odd-mode excitation 115
optimum structure 145
orthogonal resonance modes 108

oscillating circuit 104
osciplier 104, 110

parallel coupled-lines 61, 67
permeability 24
permittivity 24
phase noise 106
planar circuit 133, 146
push-push oscillator 104

Q values 24, 151
quartz crystal resonator 2
quasi-TEM 7
– mode 107, 132

radiation loss 107, 148
relative dielectric constant 20
resonance condition 13
resonator-pair 154
ring resonator 107
Rx-BPF 4

SAW 2, 5
short-circuited stub 141
single line 73
SIR 9
skin depth 26
skin effect 122
slope parameters 17
split-ring resonator 84
spurious resonance 16, 67
spurious response 7, 45, 78
spurious signals 4
spurious suppression 79
spurious-free BPF 79
step junction 12, 22
stripline 60, 73
Superconductor Filters 98
surface resistivity 26, 98

$\tan\delta_d$ 24, 46
tapered-line resonator 131
tapping 136
temperature coefficient 46
temperature stability 45
transfer function 1, 2
transmission-line resonators 7
travelling wave 108
tuned amplifier 109
two-dimensional circuits 145
Tx-BPF 4

UIR 7
undesired signals 4
unloaded-Q 24

waveguide resonators 6

Printing: Saladruck, Berlin
Binding: Buchbinderei Lüderitz & Bauer, Berlin